Concrete for Underground Structures

Concrete for Underground Structures

Editor

Rajender Patel

scitus
academics

Concrete for Underground Structures

Edited by **Rajender Patel**

Printed in 2017

ISBN: 978-1-68117-479-2

Library of Congress Control Number: 2015936595

SCITUS Academics LLC,
616, Corporate Way, Suite 2, 4766,
Valley Cottage, NY 10989

www.scitusacademics.com

Notice

Contents

Preface

Concrete is an essential tool of the underground industry and one that has a significant impact on both the durability and capital cost of a constructed project. Concrete is used in almost every project as tunnel support, backfill, or internal structure. As such, all of the parties to a project—owners, designers, and contractors—have an interest in ensuring that concrete is designed and placed in the most cost-efficient way that ensures an appropriate level of quality in the end product. Concrete is a widely used construction material for bridges, buildings and platforms as well as for underground structures such as tunnels or reinforced concrete pipelines. In general, reinforced concrete is a very durable material capable of withstanding a large range of severe environments including marine, industrial and alpine conditions.

Editor

Chapter 1

Three-Dimensional Numerical Analysis of Compound Lining in Complex Underground Surge-Shaft Structure

Juntao Chen[1, 2], Yang Yang[1, 2], Chao Ye[3], Ying Yang[1, 2], and Ming Xiao[1, 2]

[1]State Key Laboratory of Water Resources and Hydropower Engineering Science, Wuhan University, Wuhan 430072, China

[2]Key Laboratory of Rock Mechanics in Hydraulic Structural Engineering, Wuhan University, Ministry of Education, Wuhan 430072, China

[3]Wuhan Municipal Engineering Design & Research Institute Ltd., Wuhan 430015, China

ABSTRACT

The mechanical behavior of lining structure of deep-embedded cylinder surge shaft with multifork tunnel is analyzed using three-dimensional nonlinear FEM. With the elastic-plastic constitutive relations of rock mass imported and the implicit bolt element and distributed concrete cracking model adopted, a computing method of complex surge shaft is presented for the simulation of underground excavations and concrete lining cracks. In order to reflect the interaction and initial gap between rock mass and concrete lining, a three-dimensional nonlinear interface element is adopted, which can take into account both the normal and tangential characteristics. By an actual engineering computation, the distortion characteristics and stress distribution rules of the dimensional multifork surge-shaft lining structure under different behavior are revealed. The results verify the rationality and feasibility of this computation model and method and provide a new idea and reference for the complex surge-shaft design and construction.

INTRODUCTION

The surge shaft of hydropower station is a typical building in underground engineering. In recent years, with the construction of many large-scale hydropower stations, the inner diameter of surge shafts has become larger and larger, and the constructive type and geological condition have become more and more complex. According to NATM, the compound lining structure is adopted in many large-scale surge shafts, with anchoring and shotcreting support applied outside and integral lining built inside. As an important component of hydropower station structure, how to properly analyze the stability of surrounding rocks in the construction period and the structural security in the operation period is a critical technique problem.

With regard to the structural analysis theories of surge shaft, many scholars did a lot of research work [1–4]. The two most popular methods at present are traditional structural mechanics method and FEM. As to the former, the effect of surrounding rocks on lining was reflected by considering the pressure and elastic resistance of rocks, without considering the deformation compatibility and internal force adjustment in the lining-rock interaction, and the results cannot properly reflect the real stress state of structure. The latter is capable of considering the characteristics of surrounding rocks, external loads, and edge-restraint condition, and therefore it has highly forecasting precision and is applicable to the surge-shaft characteristics of multivariate bodily form and complex strained condition. Papers [5–10] use FEM to analyze the stability of surrounding rocks and operating characteristics of lining in simple formal underground structure but did not involve the analysis of complex multifork surge-shaft structure.

Based on the elastic-plastic constitutive relations of rock mass, this paper simulates the progress of excavation and lining buildup in deeply buried circular restricted orifice surge shaft by FEM, simulates the interaction and initial gap between lining and surrounding rocks with the 3D eight-node nonlinear interface element, and further proposes a calculation method for the lining cracks and structure reinforcement. In this calculation model, the main influencing factors (such as the property of surrounding rocks, the excavating and supporting process of cavern, and the interaction and initial gap between surrounding rocks and lining) of lining stress characteristics could be effectively reflected. Finally, by an actual engineering computation, the structural stability, the distortion, cracking characteristics, and the stress distribution of complex multifork surge shaft lining under different condition are revealed, which verifies the rationality and feasibility of this computing method and provides some helpful ideas and references for the structure design and construction.

NUMERICAL ANALYSIS METHOD OF SURGE-SHAFT LINING STRUCTURE

Simulation of Excavation, Support, and Lining Buildup

After the underground chambers are excavated, the stress variation in surrounding rocks is easy to cause cracks and damage. This paper describes the stress variation state of rock mass after excavation with 3D elastic-plastic damage constitutive relations [11]:

$$\left\{\sigma_{ij}\right\}_D = (1-D)\left\{\sigma_{ij}\right\} + \frac{D\left\{\sigma_{pp}\right\}\delta_{ij}}{3}, \tag{1}$$

where $\{\sigma_{ij}\}D$ is the stress of rock mass after cracking and damage, $\{\sigma_{ij}\}$ is the stress corresponding to elastic-plastic constitutive relation,

$\{\sigma_{pp}\}$ are three normal stresses, and D is the damage coefficient calculated as follows:

$$D = D_n \left[1 - \exp\left(-k\xi^a\right)\right], \tag{2}$$

where D_n, k, and a are curve fitting constants in material damage experiment and ξ^a is the 2nd invariant of rock plastic strain tensor.

The excavation load of rock mass could be divided into two parts: elastic load $\{R_e\}$ and plastic load $\{R_p\}$:

$$\{R_e\} = S\{R\}, \qquad \{R_p\} = \{1 - S)\{R\}, \tag{3}$$

where S is the structural elastic coefficient after rock excavation and the value is set to elastic coefficient of the element which firstly enters in plastic phase; that is,

$$S = \min\{r_1, r_2, \ldots, r_n\}, \tag{4}$$

where r_i is elastic coefficients of all elements that enter in plastic state after excavation. For the elastic coefficient of a certain element, theyield function F of rockmass is calculated as follows:

$$F\left(\{\sigma_0\} + r\{\Delta\sigma\}\right) = 0, \tag{5}$$

where $\{\sigma_0\}$ is the initial stress of rock mass before excavation and $\{\ \sigma\}$ is the stress incremental of rock mass after excavation

The elastic load $\{R_e\}$ is calculated once, and the plastic load $\{R_p\}$ is calculated by steps; at every step the increment load is iteratively calculated as follows:

$$[K_e]\{\Delta\delta_i\}_j = \{\Delta P_i\} + [K_D]\{\Delta\delta_i\}_{j-1}, \tag{6}$$

where $[K_e]$ is elastic rigidity matrix, $[K_D]$ is damage rigidity matrix, and the relationship of rock damage stress matrix, elastic stress matrix, and plastic stress matrix is as follows:

$$[H_D] = \left(1 - D + \frac{D\delta_{ij}}{3}\right)[H_p] + \left(D - \frac{D\delta_{ij}}{3}\right)[H_e]. \tag{7}$$

To simulate the anchor bolt and anchor cable support, the implicit anchor bolt and anchor cable element are adopted (for details see [12–14]).

The concrete lining structure is built after excavation and support of surge-shaft chamber, considering the influence of chamber deformation on lining structure; the plastic load $\{R_p\}$ released from surge-shaft excavation could be divided into two parts:

$$\{R_{p1}\} = K\{R_p\}, \qquad \{R_{p2}\} = (1-K)\{R_p\}, \tag{8}$$

where K is the excavation release coefficient decided by release property of rock mass after excavation. Before the surge-shaft lining buildup (the lining elements are still air elements), the whole structure is calculated iteratively only with the $\{R_{p1}\}$. After lining buildup (here lining elements are changed into concrete elements), the global stiffness is reformed, and then $\{R_{p2}\}$ is imposed on the whole structure for iterative computations. Finally, based on the water level diversification of surge shaft and load combinations in operating period, the deformation and stress characteristics of surge-shaft lining under different conditions are calculated; furthermore, the reinforcement ratio is figured out.

Simulation of Interface between Lining and Surrounding Rocks

The restraining effect of surrounding rocks on lining, which has great influence on the internal force of lining, could be reflected by setting interface elements [15] between lining and surrounding rocks. Based on the regular thin finite element, this paper introduces an ameliorative 3D eight-node nonlinear interface element (Figure 1) to avoid the drawbacks that only consider the gap area deformation and arbitrarily set the values of normal rigidity in Goodman interface element [16]. In constitutive relations only, with nonlinear characteristics in normal and tangential direction considered and nonlinear elastic constitutive relation imported, the interface deformation patterns, such as felt, slippage and puff, are taken into account, which can make the calculation results more actual.

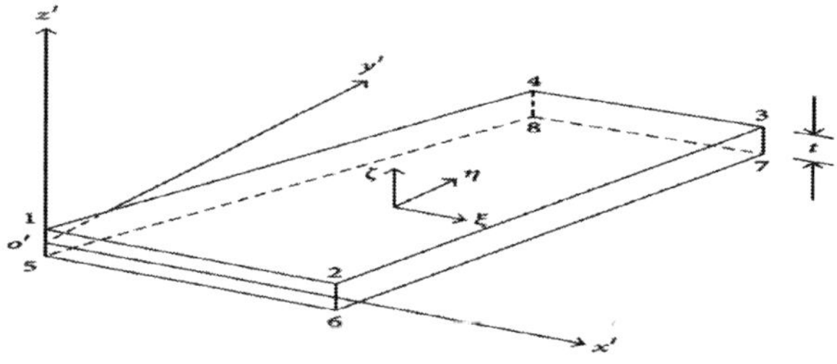

Figure 1: Nonlinear interface element.

Supposing the strain of interface element distributes evenly along thickness, the increment form of constitutive relation could be expressed as

$$
\begin{aligned}
d\{\sigma\} &= \left\{d\sigma_{x'}, d\sigma_{y'}, d\sigma_{z'}, d\tau_{x'y'}, d\tau_{y'z'}, d\tau_{z'x'}\right\}^{\mathrm{T}} \\
&= \left[D'\right]\left\{d\varepsilon_{x'}, d\varepsilon_{y'}, d\varepsilon_{z'}, d\gamma_{x'y'}, d\gamma_{y'z'}, d\gamma_{z'x'}\right\}^{\mathrm{T}},
\end{aligned}
\tag{9}
$$

where z' is the normal of interface and $[D']$ is the elastic stress matrix of interface element expressed as

$$
\left[D'\right] = \begin{bmatrix}
D_1 & D_2 & D_2 & & & \\
D_2 & D_1 & D_2 & & 0 & \\
D_2 & D_2 & k_{z'z'} & & & \\
 & & & D_3 & & \\
 & 0 & & & k_{y'z'} & \\
 & & & & & k_{z'x'}
\end{bmatrix},
\tag{10}
$$

where $D_1 = E(1 - \mu)/[(1 + \mu)(1 - 2\mu)]$, $D2 = E\mu/ [(1 + \mu) (1 - 2\mu)]$, $D3 = E/2(1 + \mu)$, E is modulus of filling material or contact material between surrounding rocks and lining, and μ is Poisson ratio.

- To describe the normal stress deformation relation of interface, the hyperbolic model of rock joint normal deformation [17, 18] is introduced:

$$k_{z'z'} = \frac{k_{z'z'0}}{(1+v/t)^2} \approx \frac{k_{z'z'0}}{(1+\varepsilon_{z'z'})^2}, \tag{11}$$

where $k_{z'z'}0$ is the initial normal rigidity of interface, its value could be set to D_1, v is normal relative deformation, and t is thickness of interface. As t is very small, $\varepsilon z'z' = v/t$. In this paper, negative value is given to the compressive stress and closing displacement of interface, with positive value given to the tensile stress and opening displacement.

- To describe the tangential stress-strains relation, the τ-u hyperbolic model [19] proposed by Clough and Duncan is adopted:

$$k_{z'x'} = \left(1 - R_f \frac{\tau'_{z'x'}}{c + \sigma'_{z'} \operatorname{tg} \varphi}\right)^2 k\gamma_W \left(\frac{\sigma'_{z'}}{P_a}\right)^n,$$

$$k_{y'z'} = \left(1 - R_f \frac{\tau'_{y'z'}}{c + \sigma'_{z'} \operatorname{tg} \varphi}\right)^2 k\gamma_W \left(\frac{\sigma'_{z'}}{P_a}\right)^n, \tag{12}$$

where τ is the shear stress, u is the tangential displacement, $k_{z'x'}$ and $k_{y'z'}$ are the tangential rigidity, c is cohesion, φ is frictional angle, γ_W is the volume weight of water, P_a is atmospheric pressure, k and n are two parameters, and R_f is the destruction ratio ($R_f < 1$).

Due to the basic principle of FEM, by the transformation relations between local coordinate system and integral coordinate system, the interface element stiffness matrix in integral coordinate system could be defined as

$$[K_e] = \int_v [B']^{\mathrm{T}} [D'] [B'] \, dv, \tag{13}$$

where $[B']$ is transformation matrix between local coordinate system and integral coordinate system.

The following strength discriminate criterions are adopted. (1) If the normal stress is greater than zero or tensile strength, the interface opens. (2) If the normal stress is less than zero or tensile strength, the Mohr-Coulomb criterion is used to discriminate whether the interface slips. During the iteration, the overloading stress occurring at the interface opening and the shear stress $\tau = \tau - (f\sigma_n + c)$ which exceeds the shear strength at the interface slipping will be transformed into the nodal loads and distributed to the surrounding elements. If the interface element crazes, the shear modulus and normal rigidity will be diminished.

Simulation of Initial Gap

Due to the quality and method of construction, the concrete shrinks when the temperature drops, the initial gaps often exist between lining and surrounding rocks, especially at the crown of chamber. Supposing the concrete lining bears all internal water pressure before the gap closes, only if the lining touches surrounding rocks after the gap closes, they both share the internal water pressure. So the initial gap can be simulated by the following methods.

- Under internal water operating condition, the interface elements set outside the lining of chamber crown is defined as initial gap elements with thickness of d (Figure 2), and their parameters are given as air elements before the gap closes. In the classification calculation process, at each step of load iteration after the nodal displacement calculation, the following progress will be cycled to all gap elements: first, it calculates the radial relative displacement u(e) of the inner face and outer face, where e is the serial number of elements. If $u(e) \geq d$, it is regarded that the initial gap is compacted and the lining touches surrounding rocks, so the initial gap elements will be changed into the compacted interface elements, and their stiffness matrix will be calculated by the method hereinbefore. If u(e) $\leq d$, no operation is required. Lastly, the loop will be closed after the judgment of all joint elements and enter the next iteration.
- Under external water operating condition, supposing that the external water pressure acts on the outer face of lining in the form of surface force, and the crown, sidewall, and base plate of

lining are in good agglomeration, so the impact of initial gap is not taken into account.

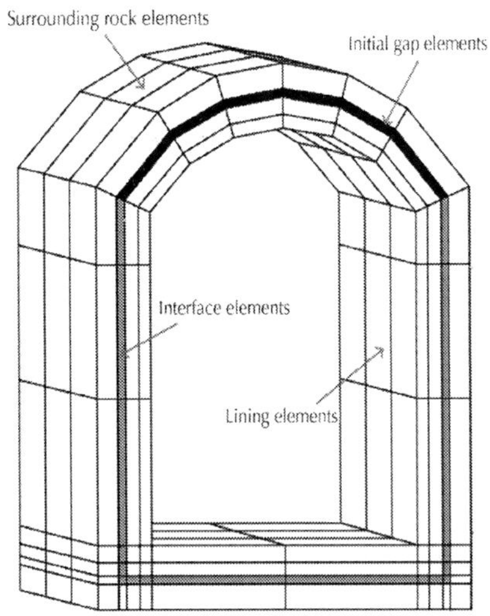

Figure 2: Simulation of initial gap.

Calculation Method of Concrete Lining Cracking

The concrete material is apt to crack under tension that causes a sudden stress variation and stiffness reduction [20]. In the iteration process, if the strain of lining elements $\varepsilon_c > \varepsilon_{d'}$ where εd is the limit tension strain of concrete, the lining element is considered as cracked. In this case, the stress $\sigma - \sigma_t - f_t$ which exceeds the concrete ultimate tensile strength f_t will be changed into the nodal loads and applied to the structure to enter the next iterative calculation. After concrete cracking, the elasticity modulus that is normal to the fracture face falls off, and the stress-strains relation changes; at this moment the crack elements are treated as anisotropic material, and the expression $\eta = 1 - D_c$ is used to describe its damage breakage properties, whereDc is the damage variable of concrete material calculated as follows [21, 22]:

$$D_c = 1 - \frac{\varepsilon_D (1 - A_T)}{\varepsilon} - \frac{A_T}{\exp[B_T (\varepsilon - \varepsilon_D)]}, \tag{14}$$

where ε is the calculation value of element principal tensile strain, ε_D is the concrete ultimate tensile strain, and A_T and B_T are two curve fitting constants; for normal concrete, $0.7 < A_T < 1$, $104 < B_T < 105$. Then the stressmatrix after concrete cracking is transformed into the matrix in global coordinate system, and the total stiffness matrix is reset to enter the next load iteration:

$$[D_{cf}] = [R]^{\mathrm{T}} [D'_{cf}] [R], \tag{15}$$

where $[R]$ is the transfer matrix between the three principal directions of elements and the global coordinate.

Based on the element strain value and structure reinforcement condition calculated by the nonlinear iteration above, the Maximal probability crack width of lining elements can be estimated as follows [23]:

$$W_{\max} = \frac{2.9\sqrt[3]{Ad}\,(1 - \rho + n\rho)\, D\varepsilon_c}{(\eta - \eta\rho + n\rho D)}, \tag{16}$$

where $A = 2rd$, d is the protection cover of reinforcement, r is spacing of bars, $n = E_s/E_c$ is the elasticity modulus ratio of reinforcement to concrete, ρ is the reinforcement ratio, and

D is the damage coefficient.

If the reinforcement is calculated by the limitative crack width, the concrete after cracking is considered as inactive; namely, the damage $D = 1$; then $\eta = 0$, and the reinforcement ratio of lining can be calculated as

$$\rho = \frac{1}{\left(1 - n + \left[W_{\max}\right] n/4t_c\varepsilon_c\right)}. \tag{17}$$

COMPUTATION MODEL AND CONDITION

Computation Model and Material Parameter

Xiaowan Hydropower Station is a state large project in China, which adopts the underground power house structure, and the installed capacity is 4200 MW (6 × 700 MW). The tailrace surge chamber uses the form of double cylindrical restricted-orifice surge shaft, three gensets share one shaft, the height of surge shaft is 90.0 m, and the diameter is 32.0 m; the lower part uses the form of complicated multifork, the tailwater tunnel is in horseshoe shape, and the diameter is 18 m. The combined form of shotcrete-bolt support and concrete lining is used in the well bore and multifork section. The chamber is deeply embedded in the lightly weathered granite gneiss, and the burial depth of number 1 surge shaft is nearly 500 m. The initial stress field of chamber excavation is simulated by 3D feedback stress field, and the three principal stresses in the center of surge shaft are −37.5 Mpa, −27.53 Mpa, and −11.49 Mpa.

Taking number 1 surge shaft and the surrounding rocks as research objects, this paper conducts computational analysis to the load-bearing characteristic of concrete lining and impedance board structure in different operating conditions. The 3D FEM mesh contains 28240 twenty-nodded isoparametric elements, which includes the tailrace surge chamber, bifurcated pipe section, impedance board, and four faults; the whole FEM mesh is shown in Figure 3. The model ranges along x-, y-, z-axes are 131.04, 168.18, and 135.00m, respectively. The thickness of lining at side wall, multifork, and impedance board is about 0.8m, 1.0m, and 1.5m; the FEM mesh of lining structure and impedance board are shown in Figure 4. The C20 concrete is adopted in lining structure, interface elements are set between surrounding rocks and lining, and the physical mechanical parameters of materials are listed in Table 1.

Table 1: Mechanical parameters of materials

Materials	Deformation modulus/GPa	Poisson ratio	C/MPa	φ /(°)	Tensile strength/ MPa	Compressive strength/MPa	Volume weight/ kN·m^{-3}
Rock Mass	18	0.25	1.2	52	1.5	80	26
Concrete	25	0.17	2.0	46	1.3	12.5	25
Interface	10	0.2	0.5	40	0.4	10	25

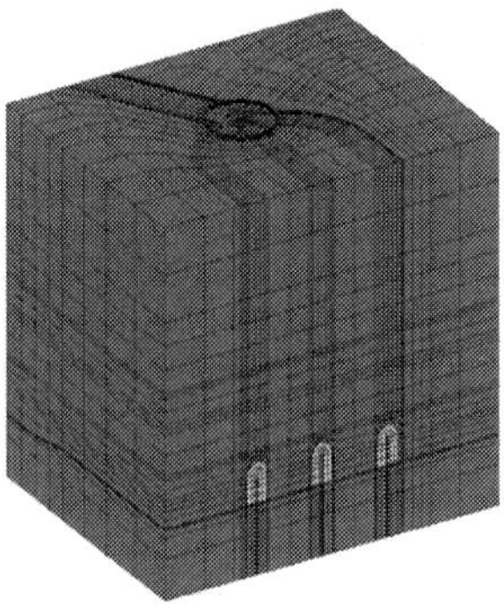

Figure 3: 3D global FEM mesh of numerical model.

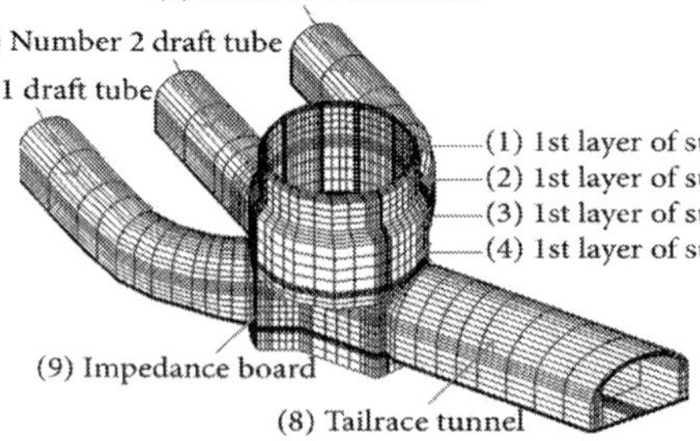

Figure 4: FEM mesh of surge-shaft lining and impedance board.

Calculation Condition

The procedure of construction including excavating, supporting and building lining is simulated. Two load conditions which are considered in FEM calculations are listed as follows.

- Normal operating condition: The main load is internal water pressure, which is calculated by the top surge water level of 1020.5 m; the acting head on impedance board is calculated by 7.15 m; the structure dead weight is included; from more secure angle, the external water pressure and rock pressure are not considered in calculation.
- Overhauling condition: The main load is external water pressure, which is calculated by the underground water level of 1033 m multiplied by reduction coefficient of 0.85; the dead weight is included; in consideration of water ramollescence, partial pressure of rocks is included; no water pressure is considered on the impedance board.

ANALYSIS OF CALCULATION RESULTS

Comparison of Calculation Results under Different Combinations of Materials

The excavation of surge chamber is simulated by excavating and supporting by stages; then lining is built and surge shaft is filled with water. In normal operating condition, two methods are used for calculation and comparison: (1) materials combination 1, considering the lining as linear elastic material without cracking and the surrounding rocks as elastic-plastic material, and (2) materials combination 2, calculating lining as nonlinear material with cracking and surrounding rocks as elastic-plastic material. The calculation results are showed in Figures 5 and 6, with positive value given to tensile axial force and inflexed bending moment.

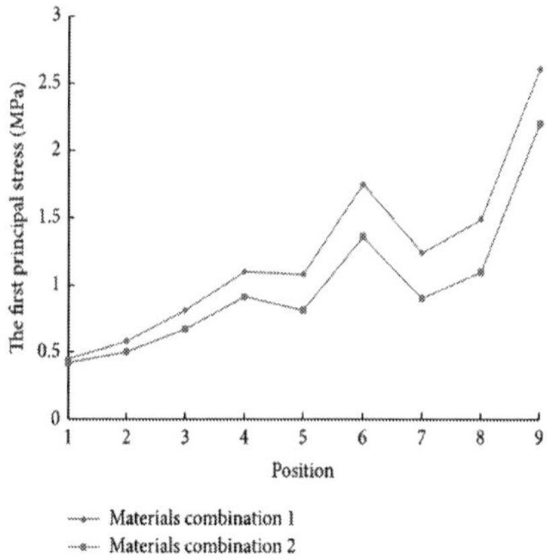

Figure 5: The first principal stress contrast figure.

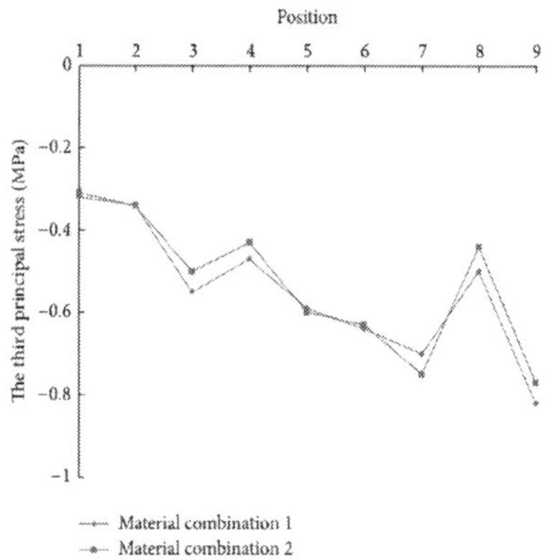

Figure 6: The third principal stress contrast figure.

By comparison of the results under internal water operating condition, the values of lining stress and internal force in material combination 1 are greater than the values in material combination 2. At the cracked place (such as the bottom of surge shaft, the intersection of chamber, the base angle of bifurcation, and top of impedance board) calculated in material combination 2, the principal stresses results of two methods differ between 16% and 27%. The results indicate that in the case of considering initial gap, while lining is regarded as linear elastic material in material combination 1, the stiffness of lining is greater than the stiffness of surrounding rocks. It is difficult to make contact and the lining will bear more load proportion after contacting, so the calculated stress values of lining and internal force are comparatively large. But in material combination 2, lining is calculated as nonlinear material with cracking. Because of the degradation of the section stiffness after lining cracking, the calculated stress and internal force are smaller than the results in material combination 1 and closer to the actual value. Overall, in the case of considering the surrounding rocks and lining as a whole and the nonlinear cracking character of lining, the calculated results are more effective to reflect the combining bearing properties of them and the actual state of stress.

Comparison of Calculation Results under Different Conditions

In the case of materials combination 2, internal water condition (normal operating condition) and external water condition (overhauling condition) are considered. The calculation results are showed in Figures 7 and8.

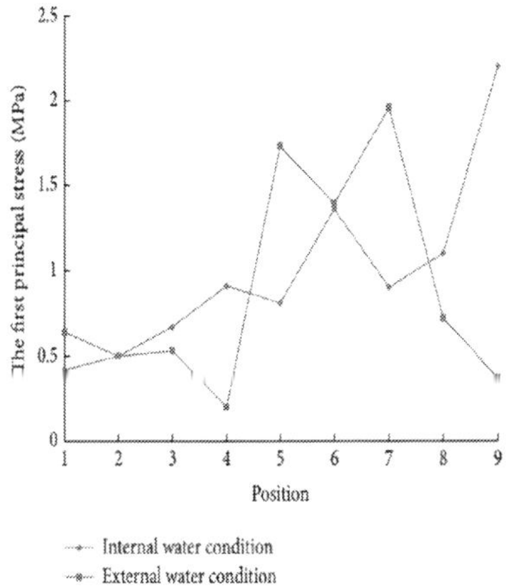

Figure 7: The first principal stress contrast figure.

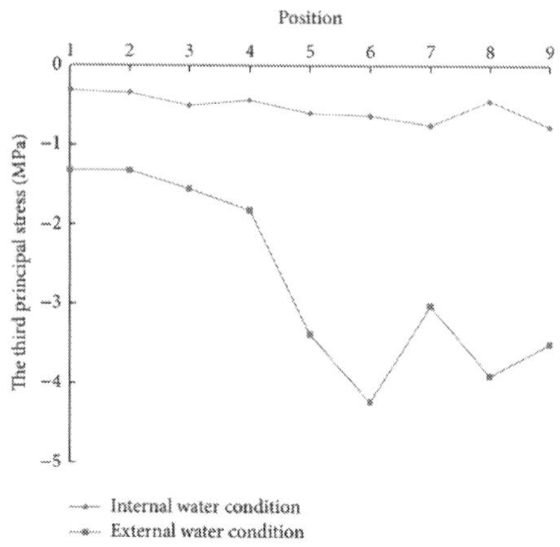

Figure 8: The third principal stress contrast figure.

Stress Characteristics of Surge-Shaft Lining and Impedance Board Structure

It can be seen that, under internal water operating condition, the stress of surge shaft lining augments gradually from top to bottom. It is chiefly because the internal water pressure increases from top to bottom of the surge shaft. The first principal stress is tangential tensile stress, with the maximal value of 2.13 MPa, which occurs at the junction of surge shaft and draft tubes (Figure 9). The third principal stress is radial compressive stress, with the maximal value of −1.07 MPa occurring at the lower part of surge shaft. The stress distributes equably in different layer of shaft. But at the variable cross-section of surge shaft and near the bifurcated pipes, affected by the junctions of chambers, the stress concentration phenomenon is obvious. The concrete showed the risk of cracking. The distribution rule of section axial force and bending moment is similar to the rule of stress; the maximum also occurs at the mid-lower part. The tensile stress at the crown and base angle of bifurcated pipes is comparatively great, which partially exceeds the tensile strength of concrete. The concrete showed the risk of cracking. The value of compressive stress is generally small, and stress concentration only occurs at the sidewall near the junctions of chambers. In general, the value of the first and third stress in most areas is not large. The structure stability of the surge shaft is guaranteed.

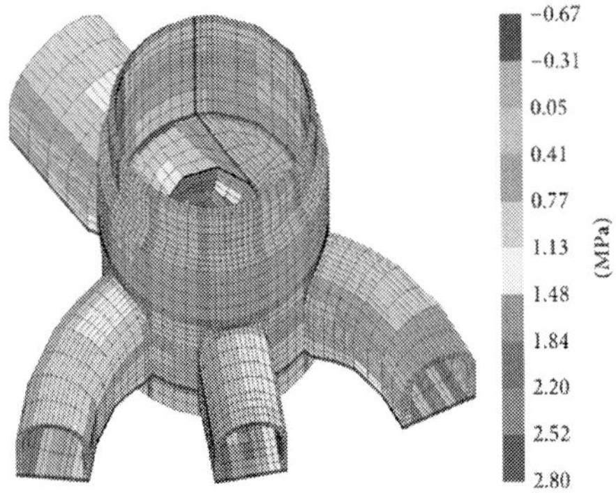

Figure 9: The first principal stress distribution under internal water pressure.

Affected by the hydrodynamic pressure at the bottom of impedance board, the tensile stress at bottom is greater than the one at top; the maximal tensile stress reaches 2.56 MPa, situated at the downstream junction of impedance board and surge-shaft sidewall. Stress concentration appears at the orifice of impedance board, and the maximal tensile stress reaches 2.2 MPa, situated at the downstream edge of orifice.

Under external water pressure of overhauling condition, the upside of surge shaft is mainly under tension, and the underside is mainly under compression, and the values of stress, internal force, and displacement are small (Figure 10). However, the tensile stress and compressive stress at the base angle of bifurcated pipes and near junctions of chambers are comparatively great. The impedance board is mainly in compression. Tensile stress occurs diminutively. The values of stress and distortion in whole impedance board are not big. This indicates that the external water pressure does not take controlling effect in the reinforcement of well bore and impedance board but only in the reinforcement of base angle of bifurcated pipes and junctions of chambers.

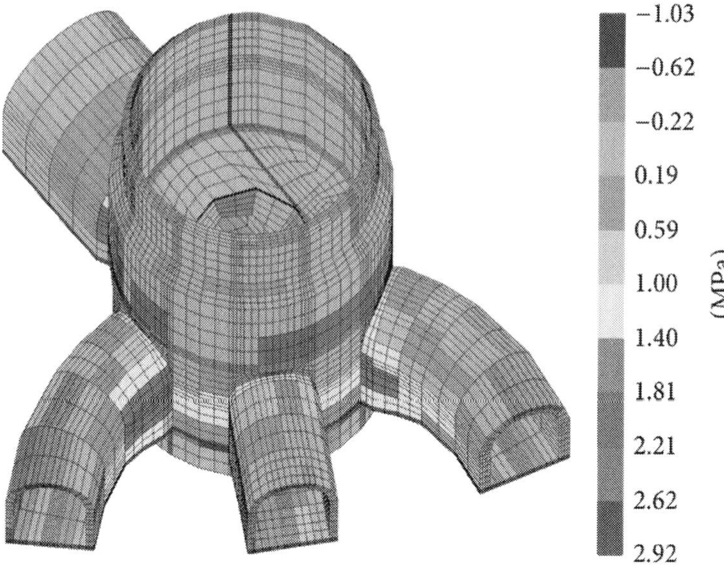

Figure 10: The first principal stress distribution under external water pressure.

Distortion Cracking Characteristics and Reinforcement Demands

Under internal water pressure, the displacement of upside surge-shaft lining is small and distributes from 0.18 mm to 0.56 mm, the displacement of downside lining is greater and distributes from 0.34 mm to 0.71 mm, and the maximal displacement occurs at the orifice of impedance board and reaches 1.49 mm. Seen from the displacement vector (Figure 11), the displacement of impedance board under hydrodynamic pressure is greater than other parts, and the lining distorts downwards and outwards under effect of deadweight and internal water pressure. It means the displacement is less affected by the constraint of surrounding rocks to sidewall.

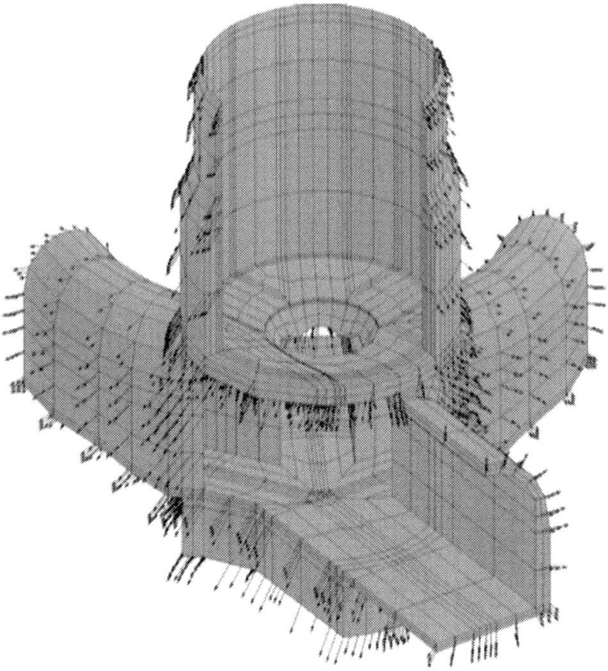

Figure 11: Displacement vector of surge-shaft lining.

The crack width and reinforcement ratio of lining in internal and external water conditions are shown in Table 2. Due to the initial gap elements set outside the crown lining, before lining contacts with surrounding rocks, the outside surface of lining is equivalent to free face; at this moment, the concrete is easy to crack under internal water pressure, and the calculated cracking width of draft tube lining and crown lining of tailrace tunnel is around 0.07 mm (Figure 12), but the cracking width of spandrel lining is smaller. At the upside of surge-shaft lining, the cracking width is small, but at the downside it is greater and distributes from 0.03 mm to 0.09 mm. The maximal cracking width is 0.15 mm and occurs at the downriver junction of impedance board and surge-shaft sidewall. Under external water effect, the structure is mainly in compression with rare cracks appearance. The calculated maximal reinforcement ratio under variable operating conditions is 1.85%, also located in the downriver junction of impedance board and surge-shaft sidewall.

Table 2: Crack width and reinforcement ratio of lining in different conditions

Operation condition	**Internal water W/ mm ρ/%**		**External water W/ mm ρ/%**	
1st layer	0.0	0.16	0.0	0.16
2nd layer	0.0	0.16	0.0	0.16
3rd layer	0.0	0.16	0.0	0.16
4th layer	0.04	0.63	0.0	0.16
Number 1 draft tube	0.07	0.91	0.0	0.16
Number 2 draft tube	0.09	1.27	0.0	0.16
Number 3 draft tube	0.08	1.19	0.0	0.16
Tailrace tunnel	0.10	1.34	0.01	0.34
Impedance board	0.15	1.85	0.03	0.51

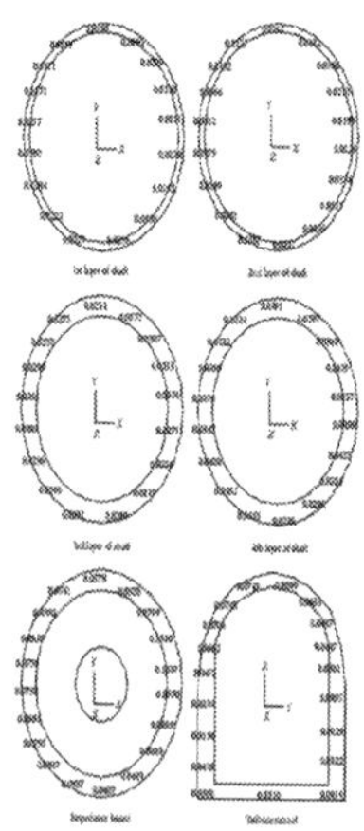

Figure 12: Crack width distribution of typical sections under internal water pressure.

CONCLUSIONS

By the 3D nonlinear FEM calculation, the deformation property and stress distribution rules of lining structure in underground surge shaft and bifurcated pipe are analyzed, and the main conclusions are drawn as follows.

- By introducing the elastic-plastic damage constitutive relations of rock mass and the implicit bolt and anchor cable element, this paper properly simulates the process of excavation, support, and lining buildup of the surge-shaft structure; the excavation and support effect of chamber could be effectively reflected. Comparison is made with the calculated results of considering concrete lining as linear elastic material and nonlinear material with cracking, which indicates that the latter is more efficient to reflect the real stress condition of lining, and the calculating method of concrete lining structure with cracking is shown to have practical significance.
- The 3D interface element including the interface normal and tangential nonlinear property is clear and definite in concept, which could effectively reflect the nonlinear relation of interface material and has good practicability. Its application in project case shows that the combined bearing and interaction between lining structure and surrounding rocks could be reasonably simulated.
- Simulation and calculated results of initial gap between lining structure and surrounding rocks show that the initial gap has great influence on the combined bearing of lining structure and surrounding rocks, the crack distribution, and width of lining structure, so good grouting treatment is necessary. Besides, the simulation method is explained to be effective to reflect the real structure stress condition and make the calculation results more practical.
- The calculation and analysis show that the internal water pressure in operating period is the primary controlling condition, and the maximal stress occurs at the orifice of impedance board and the joint between impedance board and downstream sidewall, where it should be focused on in the reinforcement designs; the base angle of bifurcated pipe and the junctions of chambers

under external water pressure should also be taken into account. Under design loading, the stress of concrete lining basically meets the material resistance requirements, the maximal crack width is in allowable range, and the deformation of the lining and impedance board basically meets the structure service demands, which shows that adopting concrete lining structure in multifork cylinder surge shaft is absolutely feasible.

ACKNOWLEDGMENTS

This study was supported by the National Key Basic Research Program of China (2011CB013506 and 2015CB057900), the Major Program of the National Natural Science Foundation of China (91215301), the National Natural Science Foundation of China (51279136 and 51209164), and the Research Fund for the Doctoral Program of Higher Education of China (20130141110015). This support is greatly acknowledged and appreciated.

REFERENCES

1. H. S. Venkatesh, R. Balachander, and R. N. Gupta, "Handling the blockades while excavating the surge shaft at Tala Hydroelectric Project," Tunnelling and Underground Space Technology, vol. 23, no. 2, pp. 145–150, 2008. · ·
2. T. Chen, X. M. Wu, and X. J. Chen, "Dynamic design and research on excavation and support design of upper surge chamber caverns of Jinping II hydropower station," Chinese Journal of Rock Mechanics and Engineering, vol. 33, no. 1, pp. 2593–2599, 2014.
3. J. A. Mello Franco, A. P. Assis, W. J. Mansur, J. C. F. Telles, and A. F. Santiago, "Design aspects of the underground structures of the Serra da Mesa hydroelectric power plant," International Journal of Rock Mechanics and Mining Sciences, vol. 34, no. 3-4, pp. 16.e1–16.e13, 1997. · ·
4. J. L. Sanchez, J. Gracia, and O. A. Fuentes, "Reliability in the design of surge tanks," International Water Power and Dam Construction, vol. 46, no. 5, pp. 36–39, 1994.

5. O. A. Murav›ev and V. V. Berlin, "Operating stability of hydroelectric stations with downstream surge tanks," Hydrotechnical Construction, vol. 29, no. 4, pp. 242–247, 1995. · ·
6. S.-A. Yufin, O.-K. Postolskaya, and E.-I. Zhokhov, "Stress state of the concrete and reinforced concrete linings of pressure tunnels in rocks with non-linear deformability," International Journal of Rock Mechanics and Mining Science, vol. 22, no. 2, pp. 56–61, 1985.
7. L. Xiang, W.-S. Zhu, Q.-S. Ma, and Q.-B. Zhang, "Damage and stability analysis of Shuangjiangkou underground caverns," Yantu Lixue/Rock and Soil Mechanics, vol. 32, no. 2, pp. 551–561, 2011.
8. J. Liu, L. Tang, J.-L. Li, J. Hu, L.-J. Fan, and J. Xia, "Stability analysis of close-located underground caverns based on tunnel lining and equivalent anchor bolt simulation," Rock and Soil Mechanics, vol. 35, no. 1, pp. 373–381, 2014.
9. J.-C. Wu, Y.-M. Zhang, and H.-X. Li, "Stability analysis of underground caverns by jointed finite element method," Applied Mechanics and Materials, vol. 238, pp. 814–817, 2012. · ·
10. M. Xiao, J.-D. Yang, and X. Lai, "Mechanics on large long corridor of surge tank," Chinese Journal of Rock Mechanics and Engineering, vol. 19, no. 4, pp. 476–480, 2000.
11. G.-W. Liu and F. Xiong, "Simulated process analysis for the excavating construction of surge-shaft,"Chinese Journal of Underground Space and Engineering, vol. 1, no. 2, pp. 305–309, 2005.
12. G. Frantziskonis and C. S. Desai, "Constitutive model with strain softening," International Journal of Solids and Structures, vol. 23, no. 6, pp. 733–768, 1987. · ·
13. J. L. Curiel-Sosa, O. A. Beg, and J. M. Murillo, "Finite element analysis of structural instability using an implicit/explicit switching technique," International Journal for Computational Methods in Engineering Science and Mechanics, vol. 14, no. 5, pp. 452–464, 2013. · · View at MathSciNet ·
14. M. Xiao, "Analysis of 3D elasto-plastic finite element for underground implicit bolt element cave,"Chinese Journal of

Geotechnical Engineering, vol. 14, no. 5, pp. 19–26, 1992.

15. M. Xiao, "Analysis of 3D nonlinear finite element for prestressed cable concrete lining," Journal of Hydraulic Engineering, vol. 30, no. 2, pp. 33–38, 1997.
16. R.-E. Goodman, R.-L. Taylor, and T.-L. Brekke, "A model for the mechanics of jointed rock," Journal of the Soil Mechanics and Foundations, vol. 194, no. 3, pp. 637–660, 1968.
17. S.-C. Bandis, Mechanical Properties of Rock Joints, Barton & Stephansson, Rotterdam, The Netherlands, 1990.
18. P.-L.-P. Wasantha, P.-G. Ranjith, A. Haque, J. Kodikara, and A. Bouazza, "Implications of joint properties on the strength of a jointed rock mass," Geotechnical Special Publication, no. 217 GSP, pp. 258–266, 2011.
19. G. W. Clough and J. M. Duncan, "Finite element analyses of retaining wall behavior," Journal of the Soil Mechanics and Foundation, vol. 12, pp. 1657–1673, 1971.
20. W. Shao, F. Jin, and G.-L. Wang, "Nonlinear thin-layer element for modeling interface," Journal of Tsinghua University, vol. 39, no. 2, pp. 34–38, 1999.
21. J.-M. Shen, C.-Z. Wang, and J.-J. Jiang, FEM of Reinforced Concrete and Limit Analysis of Plates and Shells, Tsinghua University Press, Beijing, China, 1993.
22. H.-P. Xie, Damage Mechanics of Rock and Concrete, China Mining University Press, Beijing, China, 1990.
23. J. Yuan, M. Xiao, and Z. Fu, "Numerical analysis on lining structure of restricted cylindric surge chamber in rock masses with steep obliquity," Chinese Journal of Rock Mechanics and Engineering, vol. 23, no. 8, pp. 1296–1300, 2004.

Chapter 2

Concrete Mix Design for Service Life of RC Structures under Carbonation Using Genetic Algorithm

Seung-Jun Kwon[1], Byung Jae Lee[2], and Yun Yong Kim[2]

[1]Department of Civil and Environmental Engineering, Hannam University, 133 Ojeong-dong, Daedeok-gu, Daejeon 306-791, Republic of Korea

[2]Department of Civil Engineering, Chungnam National University, 99 Daehak-ro, Yuseong-gu, Daejeon 305-764, Republic of Korea

ABSTRACT

Steel corrosion in reinforced concrete (RC) structure is such a critical problem to structural safety that many researches have been performed for maintaining required performance during intended service life. This paper is for a numerical technique for obtaining optimum concrete mix proportions through genetic algorithm (GA) for RC structures under carbonation which is considered as a serious deterioration in underground sites and big cities. For this study, mix proportions and CO_2 diffusion coefficients are analyzed through the previous studies, and then the fitness function of CO_2 diffusion coefficient is derived through regression analysis. The fitness function from 69 test results includes 5 variables of mix proportions such as w/c (water to cement) ratio, cement content, sand content percentage, coarse aggregate content, and R.H. (relative humidity). Through GA technique, simulated mix proportions are obtained for 12 cases of verification and they show reasonable results with average relative error of 4.6%. Assuming intended service life and design parameters, intended CO_2 diffusion coefficients and cement contents are determined and then related mix proportions are simulated. The proposed technique can provide initial concrete mix proportions which satisfy service life under carbonation.

INTRODUCTION

CO_2 concentration is increasing due to fossil energy consumption and this causes more carbonation damage to RC structures [1, 2]. Carbonation means that pH in pore water drops to about 10.5 due to intrusion of exterior CO_2 [3] and consumption of $CaOH_2$. In carbonated concrete, embedded steel is easily corroded. It is so critical deterioration phenomenon that it should be considered in durability design for underground RC structures or those in metropolitan cities which have high CO_2 concentration.

With higher CO_2 concentration, carbonation depth increases but this can be comparatively controlled by a design of concrete mix proportions. The influencing parameters on carbonation are reported to be type of cement, unit content of cement, type of aggregate, and so on [2]. Semiempirical prediction techniques, so-called mesolevel, have been proposed and they are still utilized for the sake of simple and practical application [4, 5]. Carbonation mechanism can be explained as diffusion of CO_2 and carbonatable materials like calcium hydroxide ($Ca(OH)_2$) and calcium silicate hydrates (C–H–S). CO_2 diffusion represents how fast CO_2 gas (or liquid) intrudes into concrete, so that concrete with high CO_2 diffusion coefficient allows rapid carbonation. From the defensive point of view for carbonation, concrete with larger carbonatable materials can keep high alkali so long as they are not fully consumed due to carbonation reaction. From 1980, several physico-chemo carbonation models have been proposed. They are all constructed by both modeling on diffusion coefficient based on pore structure and modeling on carbonic reaction based on dissociation of carbonatable materials [3,6–8]. Carbonation modeling for cracked and joint concrete is similarly performed considering the larger CO_2intrusion due to crack effect and cold joint effect [9–12]. Recently, carbonation prediction techniques are proposed through experimentally measuring CO_2 diffusion coefficient [13–15] and numerically obtaining CO_2diffusion coefficient through neural network algorithm [16].

If environmental conditions like CO_2 concentration, temperature, and R.H. are quantitatively evaluated, intended carbonation depth in design stage can be determined considering design cover depth and intended service life. Provided that various mix proportions and the related CO_2 diffusion coefficients are experimentally given, intended CO_2 diffusion coefficient satisfying the intended service life can be obtained. Then mix proportions satisfying the intended CO_2 diffusion coefficient can be obtained through optimization technique as well.

GA (generic algorithm) technique is a representative optimization technique and widely utilized in civil engineering. Through reverse analysis, the parameters which satisfy the fitness function can be derived so that application of GA has been extended. For the application of GA to concrete researches, mix proportion optimizations are performed only for strength prediction in HPC (high performance concrete) [17, 18]. With regard to durability design for service life, very limited research has been performed for chloride attack [19]. For carbonation, optimization of mix proportions has not been carried out so far.

In this paper, CO_2 diffusion coefficients and the related mix proportions are investigated. Based on 69 mix proportions and diffusion coefficients; the fitness function for CO_2 is derived through MATLAB with parameters of mixing variables (w/c ratio, unit content of cement, and fine and coarse aggregates) and exterior variables (R.H.). Through comparison with the previous test results, the applicability of GA technique for optimum mix proportions is verified. Assuming the intended service life and environmental conditions, the intended CO_2 diffusion coefficient is calculated. Finally, the mix proportions which satisfy the intended CO_2 diffusion coefficient are derived through GA technique. This technique can be utilized for performance-based concrete mix design. The techniques for carbonation prediction and optimization of mix proportions are dealt with in this paper.

BACKGROUND OF GA AND INFLUENCING PARAMETERS ON CARBONATION

Overview of GA

Unlike conventional search technique, GA technique constructs arbitrary solutions in initial group, and then the fittest solution is derived through modification of the solutions. GA technique is mainly utilized in the field of mechanical and electrical engineering and recently applied to civil engineering such as design optimization for structures, line network analysis, and concrete mix design for strength. This technique can provide more accurate results than other algorithms having many local solutions [20, 21]. GA technique starts with an initial set of random solutions called population. Each individual in the population is called a chromosome, which represents a solution to the problem at hand. The evolution operator simulates Darwinian evolution process to create population from generation to generation. The availability of genetic algorithm depends on its ability to keep existing parts of solution, which have a positive effect on the outcome, and proceed with optimizing the no optimal part. The transition rules which combine and change those samples for improving the solutions are probabilistic and not deterministic. This enables genetic algorithm to reach a global optimum without being fixed in local optima [18]. In the selection stage, GA fundamentally starts from Darwinian natural selection and the initial individuals are selected in this process. Selection provides the driving force in genetic algorithm, and selection pressure is critical in it. The selection directs genetic algorithm search toward promising regions in the search space [18]. The second stage, crossover is the most important genetic operator in which the bit-strings of two (or more) parents are cut into two (or more) pieces and the parts of bit-string are crossed over. The point where the parents are cut is randomly determined. Through the crossover operator, a new child population has been created using inherited values from the parent population. Mutation operator is used to insert new information into the new population, preventing GA from getting stuck in certain regions of the parameter space [18]. Mutation consists of making slight changes in parameters of child population after they have been generated by crossover. More detailed information on GA can be found in many researches [17–21]. The process of GA is presented in Figure 1.

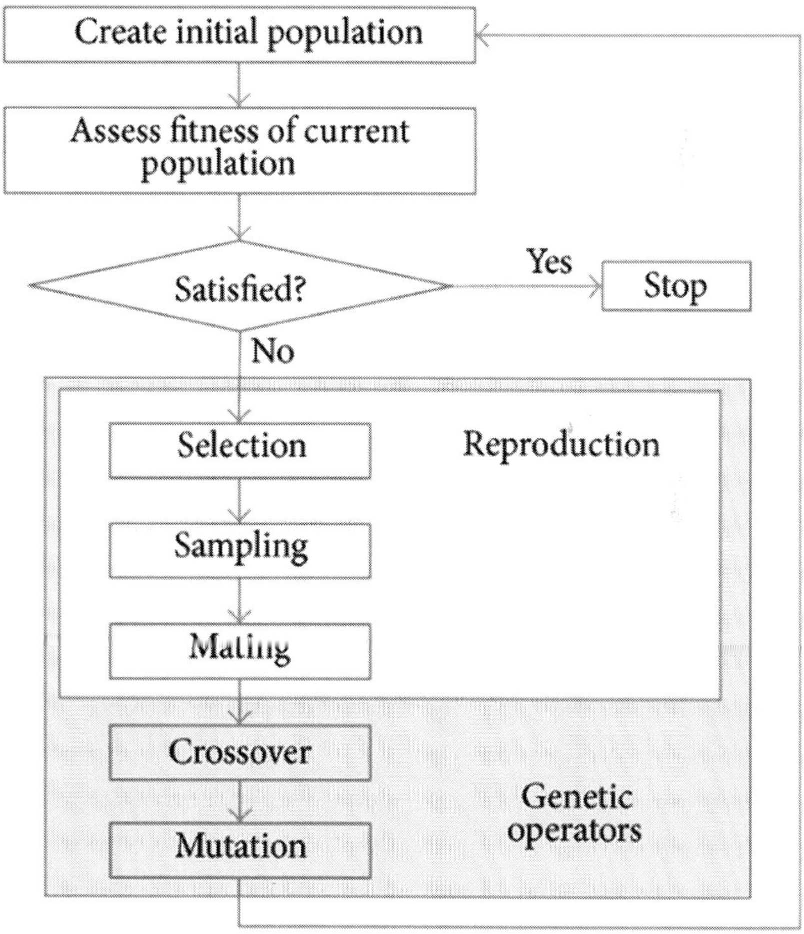

Figure 1: Genetic algorithm process [29].

Study of Carbonation Parameters and Prediction Techniques

Influencing parameters on carbonation can be classified into two groups. One is for external parameters regarding environmental conditions and the other is for internal parameters regarding diffusion coefficient and carbon table materials. Considering these parameters, many carbonation prediction techniques have been proposed in semi empirical form. These equations assume that carbonation depth is proportional to square root of exposed time. This assumption was verified through experiments, field investigations [1, 2, 9, 22], and analytical solution [6, 23]. In Table 1, carbonation parameters are summarized. Conventional techniques for carbonation prediction are listed in Table 2 [1, 2].

Table 1: Influencing parameter on carbonation behavior

Internal parameter (mixture)	Low w/c and large unit cement amount	(i) Holding pH in alkali through producing large amount of hydration of CSH and $Ca(OH)_2$ (ii) Low CO_2 diffusion through dense pore structure
	Aggregate	CO_2 intrusion through artificial light weight aggregate
	Mineral admixture (slag and fly ash)	(i) Small amount of $Ca(OH)_2$ due to pozzolanic reaction and latent hydraulic reaction
		(ii) Low diffusion coefficient of CO_2
	Mixed chloride content	Rapid carbonic reaction due to high pH from ion dissociation
	Alkali	(i) Rapid carbonic reaction due to high alkali cement
		(ii) Residual metallic oxide (K_2O, Na_2O)

External parameter	CO_2 concentration	Rapid carbonation through higher concentration of CO_2
	Temp.	Increasing activity energy due to high temperature (Arrhenius law)
	R.H.	(i) Decreasing carbonation in low R.H. due to insufficient H_2O
		(ii) Decreasing carbonation in high R.H. due to low CO_2 diffusion
	Induced chloride ion	Rapid carbonation due to dissociated chloride ion (cation)

Table 2: Semiempirical equations for carbonation process

Researcher	Equations
Syrayama	$t = \alpha\beta\gamma\delta\varepsilon\dfrac{5000C^2}{(x-38)^2}$
	C: carbonation depth,x: w/c ratio
	, , , , : factors for admixtures, cement type, exposure condition, and so forth
Kishitani	$t = \dfrac{0.3(1.15+3\mathrm{w/c})C^2}{R^2(\mathrm{w/c}-0.25)^2}$ (w/c $\geq$ 0.6)
	$t = \dfrac{7.2C^2}{R^2(4.6\mathrm{w/c}-1.76)^2}$ (w/c $<$ 0.6)
	C: carbonation depth
	R: factor for cement type, aggregate type, and surface treatment
Hamada	$t = \dfrac{kC^2}{R}, \quad k = \dfrac{0.3(1.15+3x)}{(x-0.25)^2}$
	C: carbonation depth,x: w/c ratio
	R: factor for cement type, aggregate type, and surface treatment

Ida	$t = \alpha\beta\gamma\frac{KC^2}{(100x-18)^2}$
	C: carbonation depth,x: w/c ratio
	K: factor for exposure and cement type
	, , : quality, retardation, and environmental condition

The flowchart for this study is shown in Figure 2. Through this study, fitness function for diffusion coefficient, intended diffusion coefficient for service life, and mix proportions satisfying intended diffusion coefficient are derived using GA technique.

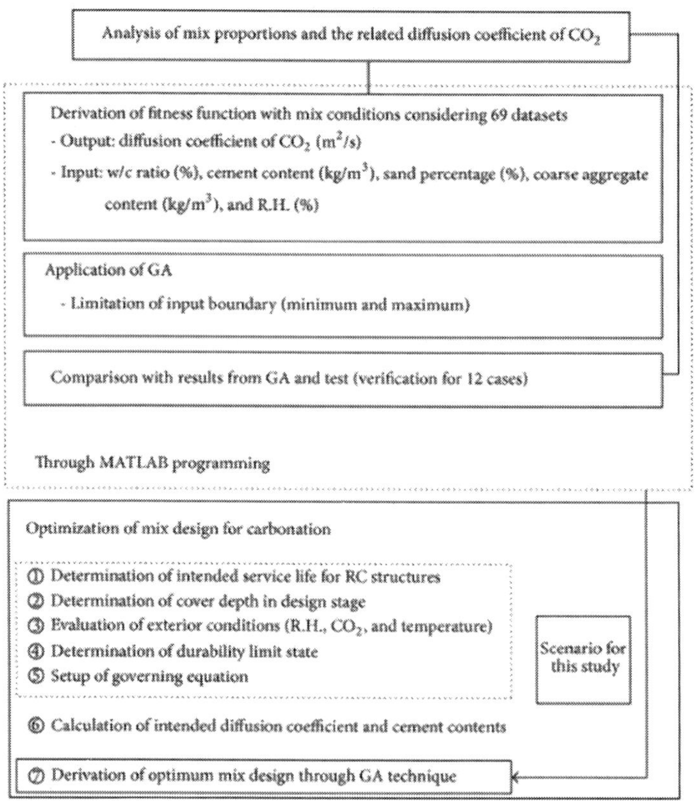

Figure 2: Flowchart for this concrete mix optimization.

CONCRETE MIX OPTIMIZATION USING GA

Fitness Function for Diffusion Coefficient

Previous Test for Diffusion of CO_2 [14, 16]

For the derivation of fitness function, the previous test results are adopted. In the test, CO_2 diffusion coefficients were measured through diffusion cell. Three different mix conditions and 4 different R.H. were considered [14]. So far, several researches have been reported for the test of CO_2 diffusion coefficients; however, they are not for concrete but for cement mortar or cement paste [6, 13]. Very limited cases are reported for CO_2 diffusion coefficient in concrete. In Table 3, the procedures for the adopted test are summarized. Mix proportions and cement properties are listed in Table 4. The test adopted in this paper covers only OPC (Ordinary Portland Cement, type I) concrete since concrete with mineral admixtures has different carbonation behavior due to the decreased diffusion coefficient and pozzolan reaction [3, 7].

Table 3: Summary of test setup [14]

Steps
(a) Installation of test equipment in room (20°C)
(b) Measurement of concrete sample thickness and diameter
(c) Installation of sample (concrete disk) in cell
(d) Applying N2 gas and CO2 gas to different cells with same pressure
(e) Measurement of CO2 concentration when CO2 concentration in N2 gas keeps constant (steady state) $D_{CO_2} = \dfrac{Q f_{CO_2} L}{(1 - f_{CO_2}) A}$

D_{CO_2} : diffusion coefficient of CO2;Q: flow rate of gas
f_{CO_2} : mol fraction in N2 + CO2;L: thickness of disk;A: area of disk

Table 4: Mix proportions for CO_2 diffusion measurement [14]

(a)					
Case	w/c (%)	Cement (kg/m3)	Water (kg/m3)	Sand (kg/m3)	Coarse aggregate (kg/m3)
1	42	425	179	714	895
2	50	315	158	748	1,076
3	58	277	161	726	1,117

(b)			
Aggregate properties			
Type	Specific gravity	Absorption (%)	Fineness modulus
Fine	2.56	2.18	2.85
Coarse	2.60	0.94	6.51

Derivation of Fitness Function

The adopted test was performed considering 4 different R.H. as 10%, 45%, 75%, and 90%. In order to obtain more reasonable fitness function, several previous test results [15, 24, 25] are considered. Carbonation process is very sensitive to R.H. since concrete with high saturation allows active carbonation reaction but low diffusion of CO_2, and concrete with low saturation allows high diffusion of CO_2 but it has little H_2O for carbonic reaction. With higher R.H. and lower w/c ratio, CO_2 diffusion coefficients decrease as in Figure 3. In Figure 3, several results [25, 26] are obtained from reverse analysis based on measured carbonation depth with constant R.H.

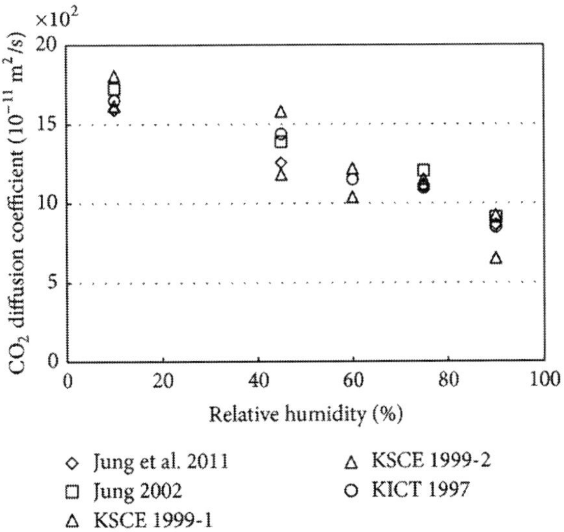
×10^2
CO2 diffusion coefficient (10^-11 m^2/s)
Relative humidity (%)
Jung et al. 2011
Jung 2002
KSCE 1999-1
KSCE 1999-2
KICT 1997

(a)

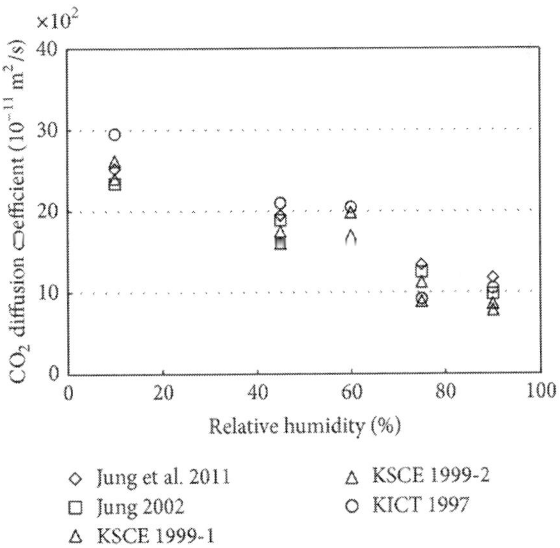
×10^2
CO2 diffusion coefficient (10^-11 m^2/s)
Relative humidity (%)
Jung et al. 2011
Jung 2002
KSCE 1999-1
KSCE 1999-2
KICT 1997

(b)

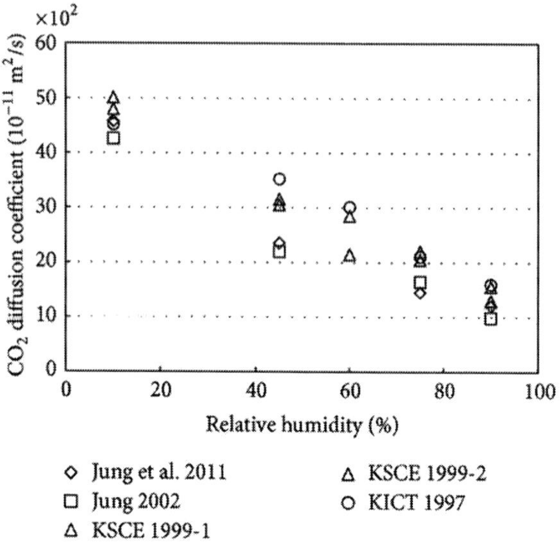

(c)

Figure 3: CO_2 diffusion coefficients with w/c ratios and R.H.

For the relation with mix proportions and CO_2 diffusion coefficient, fitness function with mix components should be obtained. In the previous researches, fitness function for strength was derived through linear multiregression curve, which contained the variables of mix components like w/c ratio and unit amount of cement [18]. Unlike the fitness function for strength, R.H. is very critical to CO_2 diffusion coefficient, so that both mix components (w/c ratio, content of cement, sand ratio, and content of coarse aggregate) and R.H. are considered as variables in the fitness function in this analysis.

In the optimization technique, many local solutions can be obtained. For avoiding convergence to local solution, initial variables (starting variables) and wide ranges for each solution are necessary. Even wide ranges of solutions are considered, local solutions may be obtained because of the initial variables in conventional optimization techniques, so that GA technique is preferred for searching solution in overall ranges. The variables and the related ranges are listed in Table 5.

Table 5: Variables for fitness function

Type	w/c (%)	C (kg/m3)	S/a (%)	G (kg/ m3)	R.H. (kg/ m3)	a and b	Constant
Max	100	0	10	10	30	1	200,000
Min	−100	−100	−10	−10	−30	−1	0

With larger unit content of cement, diffusion coefficient decreases, so that C in Table 5 is set to have below zero. w/c ratio is assumed to have a range of −100~100. S/a (sand to total aggregate) and G have relatively small effect on CO_2 diffusion, so that they are assumed to have small range of −10~10. CO_2 diffusion coefficient is much dependent on R.H., so that the range of R.H. is assumed as −30~30.

Typical multiregression analysis is shown in (1). In (2), additional term for the consideration of R.H. is added. Averages of relative error are evaluated to be 17.3% from (1) and 7.6% from (2), respectively. The results of multiregression curves are listed in Table 6. For the derivation of constant in (1) and (2), GA technique is utilized. Consider

$$D_{CO_2} = I + A\left(\frac{w}{c}\right) + B(C) + C\left(\frac{S}{a}\right) + D(\text{Agg}) + E(\text{RH}), \quad (1)$$

$$D_{CO_2} = \left[I + A\left(\frac{w}{c}\right) + B(C) + C\left(\frac{S}{a}\right) + D(\text{Agg}) + E(\text{RH})\right](a\text{RH} + b). \quad (2)$$

Table 6: Variables and constants in regression analysis

	I	A	B	C	D	E
Equation (1)	1018.56	16.08	−1.42	2.75	0.60	−9.75

Equation (2)	15,427.82	49.562	−29.60	5.48	−4.42	9.79
	a: −0.005825;b: 0.7542					

w/c: w/c ratio (%); C: cement content (kg/m^3); S/a: sand percentage (sand/total aggregate) (%); Agg: coarse aggregate content (kg/m^3); R.H: relative humidity (%); *I*: Intersection (constant).

In Figure 4, the results of regression analysis ((1) and (2)) and test results (averages) are compared. As shown in Figure 4, when (1) is selected, it provides a big error for the case of high w/c ratio (w/c 58%), so that (2) is selected for fitness function for this study. For the case of w/c 58%, (1) shows the relative error range of 10.6~24.3% but (2) shows 3.3~14.2%. If a better fitness curve can be defined through nonlinear regression analysis, it would provide the more reasonable mix proportions based on the test dataset.

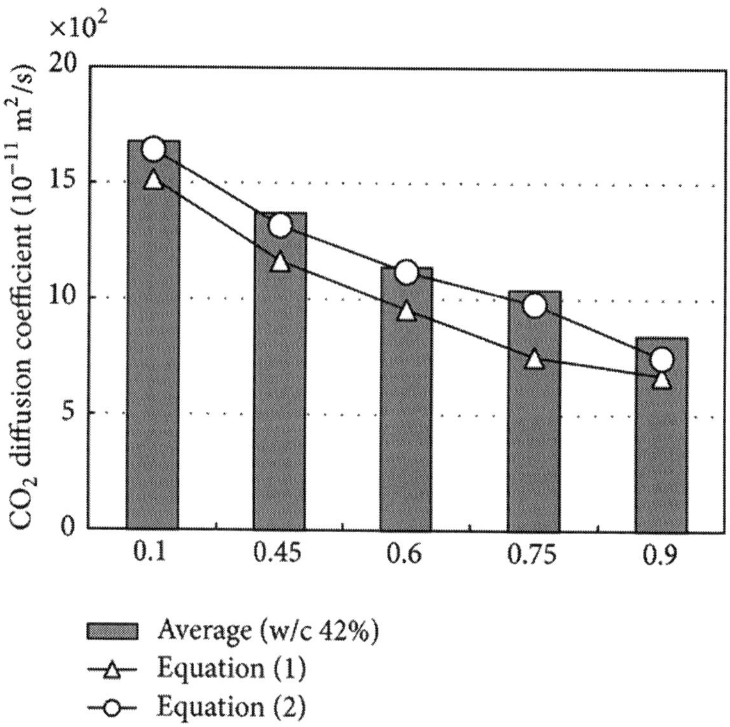

(a)

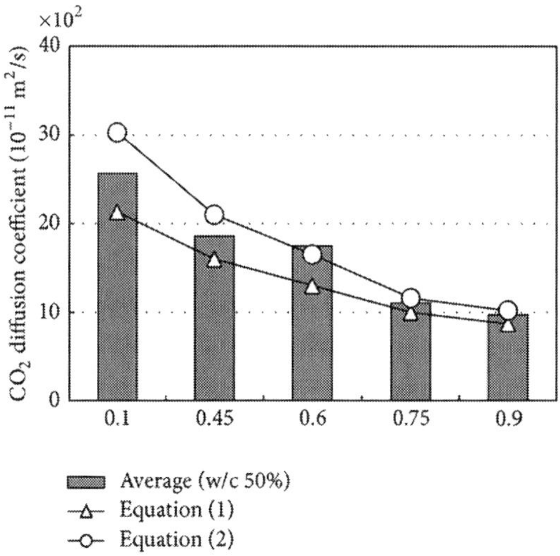

(b)

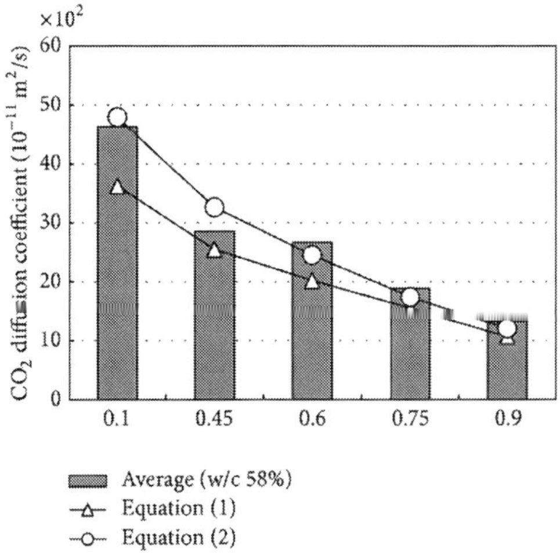

(c)

Figure 4: Test and simulated results for diffusion coefficient.

CO_2 diffusion coefficients are strongly dependent on mix proportions. The contours are shown in Figure 5. Case 1 represents w/c 42% in Table 4, where unit cement content is changed from 277 kg/m^3 to 425 kg/m^3 and R.H. is changed from 10% to 90%. Cases 2 and 3 show the simulations of CO_2 diffusion coefficient in w/c 50% and 58%, respectively With larger cement content and higher R.H., CO_2 diffusion coefficients decrease in every case.

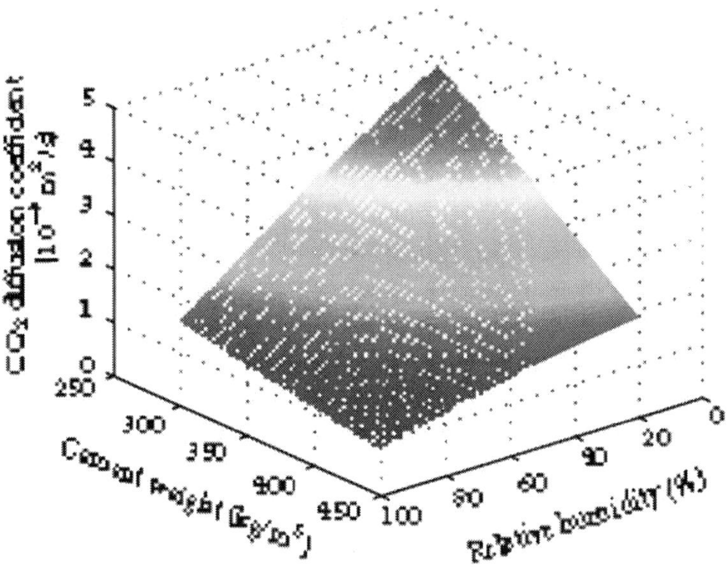

(a)

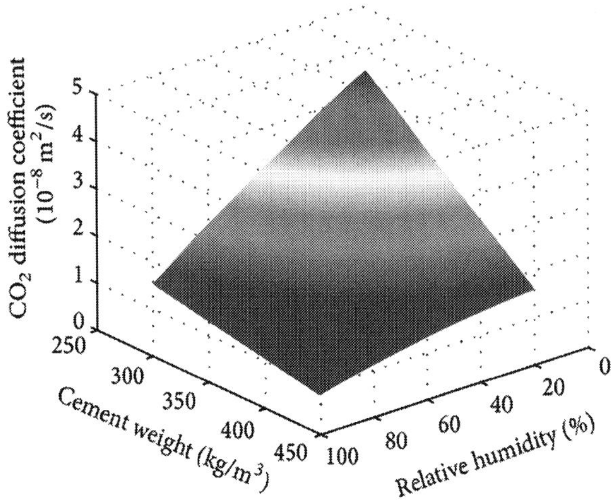

(b)

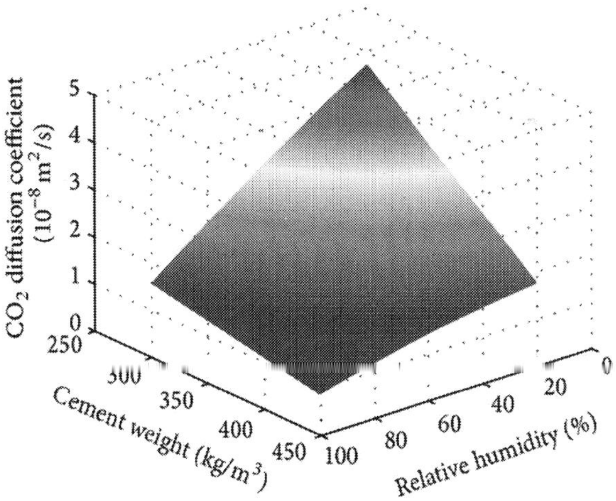

(c)

Figure 5: Diffusion coefficient contour with cement content and relative humidity.

Evaluation of GA Applicability to Generating Mix Proportion

In order to evaluate the applicability of GA, verification is performed for 3 different cases (w/c 42% with R.H. 10%, w/c 50% with R.H. 75%, and w/c 58% with R.H. 90%). Population size is set as 20 and the number of generation is set as 10,000 for avoiding early convergence. For formation of 1st generation, uniform function is adopted. For parent selection for next generation, stochastic uniform function is utilized and two superior chromosomes are transferred to next generation. Crossover function of two-point is adopted and normal distribution is considered for mutation operator with mutation ratio of 0.8.

Determination of up/down boundary conditions is important to obtain each mix component and this needs user's experience. The range of boundary conditions, obtained mix components through GA, and the range of relative errors are listed in Table 7. Output results are mix component and CO_2 diffusion coefficient. R.H. is set to fix since it can be known from exterior condition. The fixed R.H. is made through letting up/down boundaries have the same R.H. value.

Table 7: Comparison with results from test and simulation from GA

w/c (%)	Diffusion coefficient (10–11 m2/ sec)	w/c (%)	R.H. (%)	Cement	S/a (%)	Coarse aggregate (kg/m3)
42	1574	42.0	10	425	44.4	895
Input range	—	40–45	10-10	400–450	39–45	800–895
Result from GA	1580.5	41.5	—	419.2	41.8	874.5
Relative error (%)	0.4	−1.2		−1.4	−5.9	−2.3
42	1257	42.0	45	425	44.4	895
Input range	—	40–45	45-45	400–450	39–45	800–895

Result from GA	1194.2	42.1	—	420.8	42.2	869.7
Relative error (%)	−5.0	0.24		0.0	−5.0	−2.9
42	1105	42.0	75	425	44.4	895
Input range	—	40–45	75-75	400–450	39–45	800–895
Result from GA	1088.4	42.2	—	421.2	42.0	882.1
Relative error (%)	−1.5	0.5		−0.9	−5.4	−1.4
42	862	42.0	90	425	44.4	895
Input range	—	40–45	90-90	400–450	39–45	800–895
Result from GA	855.2	41.8	—	426.1	43.2	887.2
Relative error (%)	−0.8	−0.5		0.3	−2.7	−0.9
50	2520	50.0	10	315	41.0	1076
Input range	—	47.5–52.5	10-10	290–330	39–45	950–1100
Result from GA	2775.2	49.8	—	311.8	42.6	1085.3
Relative error (%)	10.1	−0.4		−1.0	3.9	0.9
50	1950	50.0	45	315	41.0	1076
Input range	—	47.5–52.5	45-45	290–330	39–45	950–1100
Result from GA	2124.2	51.3	—	318.5	41.2	1092.5
Relative error (%)	8.9	2.6		1.1	0.5	1.5

50	1503	50.0	75	315	41.0	1076
Input range	—	47.5–52.5	75-75	290–330	39–45	950–1100
Result from GA	1452.2	49.3	—	322.2	41.8	1044.7
Relative error (%)	−3.4	−1.4		2.3	2.0	−2.9
50	1105	50.0	90	315	41.0	1076
Input range	—	47.5–52.5	90-90	290–330	39–45	950–1100
Result from GA	1127.3	48.2	—	318.9	39.5	1068.5
Relative error (%)	4.0	−3.6		1.2	−3.7	−0.7
58	4480	58.0	10	277	39.4	1117
Input range	—	55–60	10-10	260–330	39–45	950–1300
Result from GA	4922.3	57.2	—	266.2	39.7	1204.2
Relative error (%)	9.7	−1.4		−3.9	0.8	7.8
58	2350	58.0	45	277	39.4	1117
Input range	—	55–60	45-45	260–330	39–45	950–1300
Result from GA	2472.3	58.9	—	270.5	40.1	1200.8
Relative error (%)	5.2	1.6		−2.4	1.8	7.5
58	1450	58.0	75	277	39.4	1117
Input range	—	55–60	75-75	260–330	39–45	950–1300

Result from GA	1377.2	57.8	—	277.7	42.2	1208.7
Relative error (%)	−5.0	−0.3		0.3	7.1	8.2
58	1172	58.0	90	277	39.4	1117
Input range	—	55–60	90-90	260–330	39–45	950–1300
Result from GA	1150.2	59.4	—	278.5	41.9	1187.5
Relative error (%)	−0.9	2.4		0.5	6.4	6.3

For 12 data [14], the comparison with test and simulated results from GA are shown in Figure 6(a) with regard to CO_2 diffusion coefficient, which shows a reasonable agreement. In Figure 6(b), the comparison of relative errors is shown for 12 cases and the average relative error of each component is shown in Figure 6(c). The processes of searching the optimum solutions are plotted in the case of w/c 42% with R.H. 10% from Figure7 (a) to Figure 7(e).

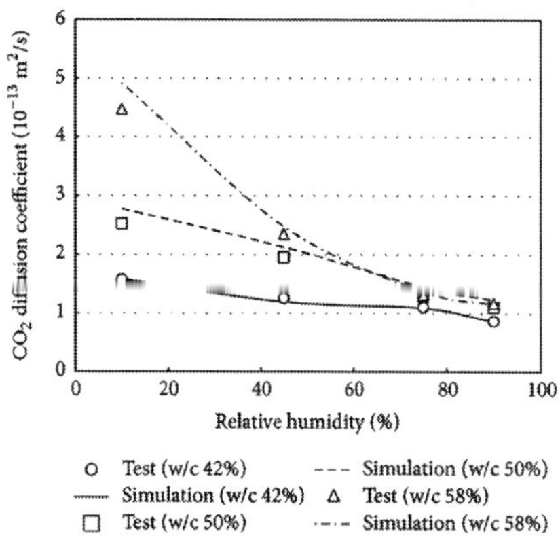

(a)

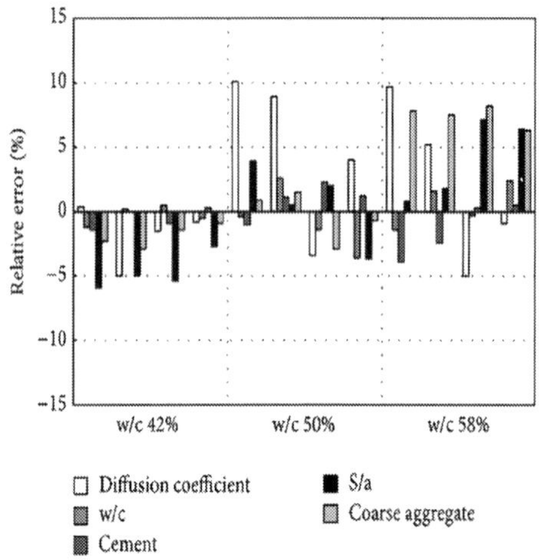

(b)

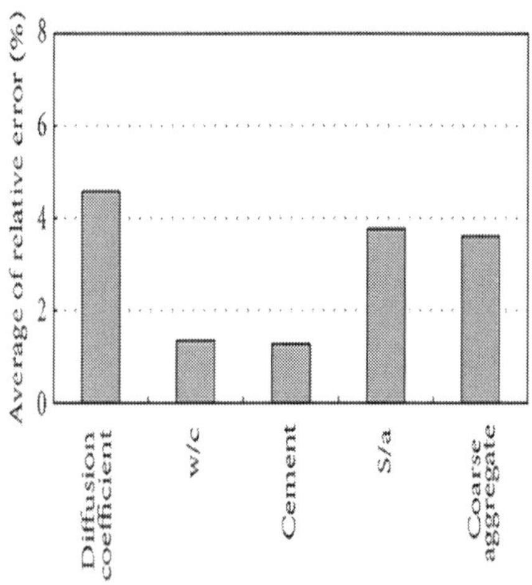

(c)

Figure 6: Comparison with results of CO_2 diffusion coefficient and relative errors.

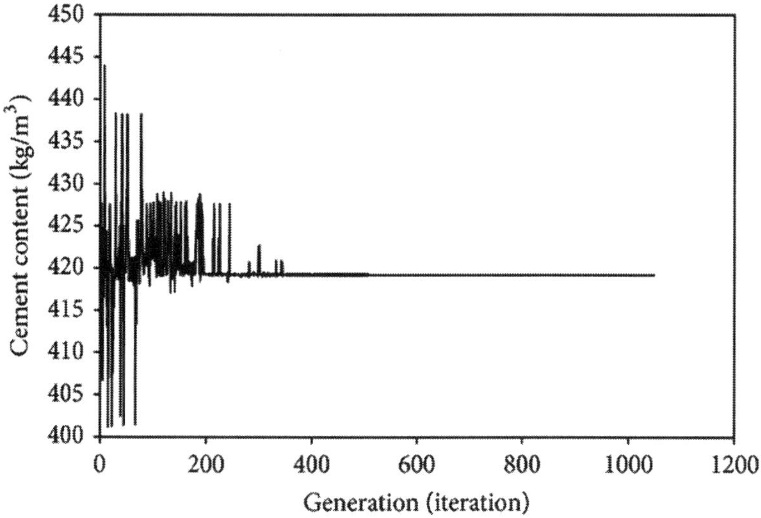
450
445
440
435
430
425
420
415
410
405
400
Cement content (kg/m³)
0
200
400
600
800
1000
1200
Generation (iteration)

(a)

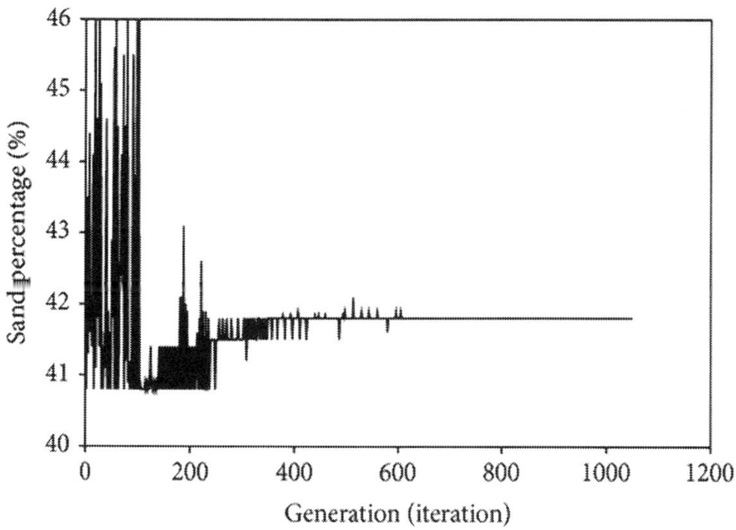
46
45
44
43
42
41
40
Sand percentage (%)
0
200
400
600
800
1000
1200
Generation (iteration)

(b)

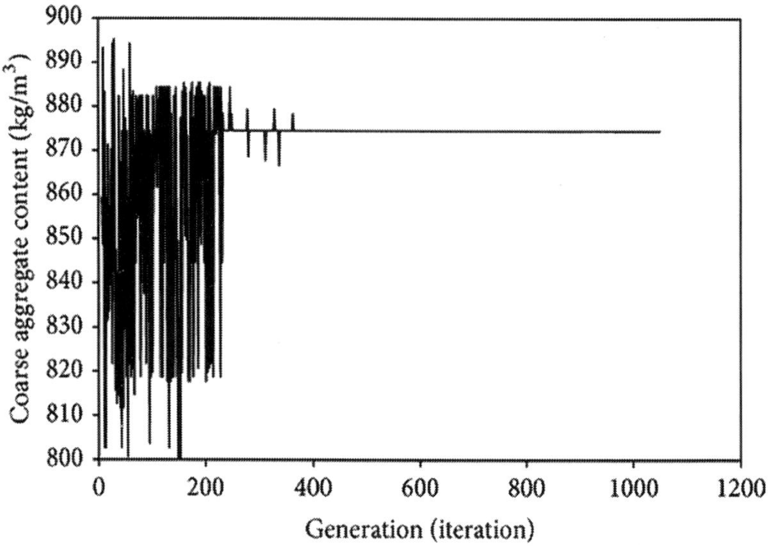
900
890
880
870
860
850
840
830
820
810
800
Coarse aggregate content (kg/m3)
0
200
400
600
800
1000
1200
Generation (iteration)

(c)

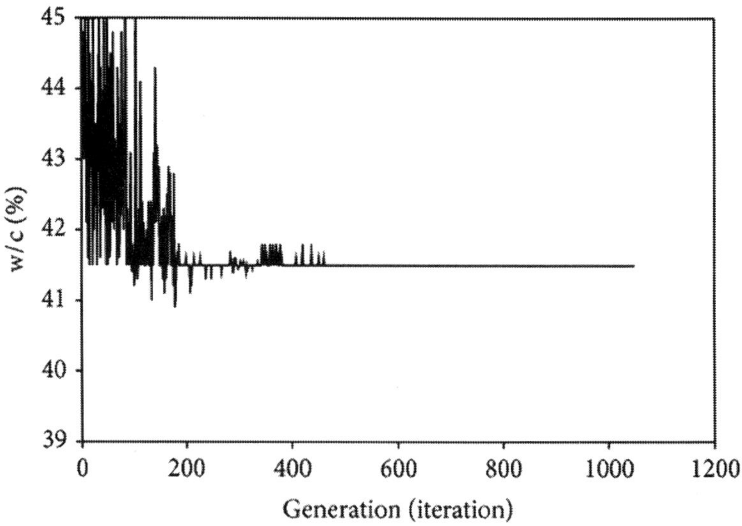
45
44
43
42
41
40
39
w/c (%)
0
200
400
600
800
1000
1200
Generation (iteration)

(d)

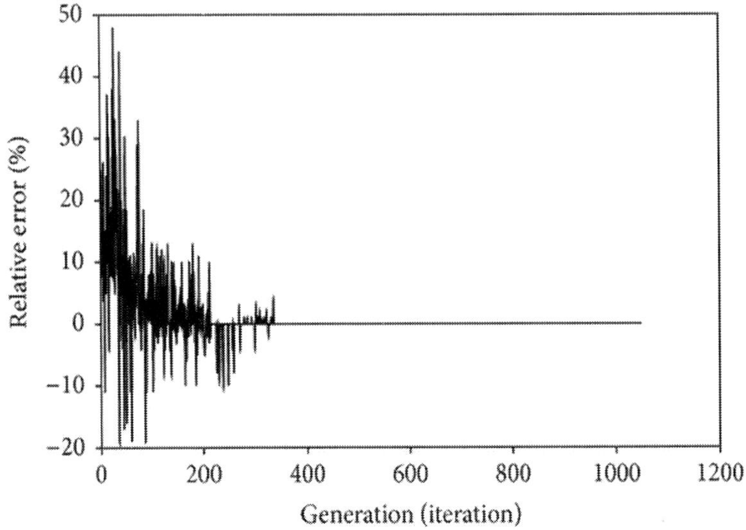

(e)

Figure 7: Simulated process through GA (w/c 0.42 and R.H. 10%).

As listed in Table 7, this technique reasonably estimates the CO_2 diffusion coefficients and mix proportions with −5.0~10.1% of relative errors.

DESIGN OF CONCRETE MIX PROPORTIONS FOR CARBONATION

Scenario for Mix Design Considering Carbonation

In this section, concrete mix design is performed considering exterior condition-carbonation. The design flow is as follows:

- determination of intended service life,
- determination of design cover depth,

- evaluation of exterior condition,
- determination of durability criteria,
- determination of governing equation,
- Mix optimization through GA.

If reduction factors or safety factors are considered [26, 27], conservative design can be induced. However, intended diffusion coefficient is derived assuming 1.0 of reduction and safety factor in this paper. Generally, underground site and urban area are reported to be the environments where durability design for carbonation is necessary since they have relatively high CO_2 concentration and normal R.H. (50%~70%). In the previous research [25], durability design for carbonation is strongly recommended over 300 ppm of CO_2 concentration. In urban cities, CO_2 concentration over 350 ppm is reported; furthermore, CO_2 concentration over 650 ppm is reported in underground sites like subway structures [25]. Several specifications [26–28] guide durability design for carbonation in urban cities and underground structures.

Mix Design Considering Exterior Conditions and Design Parameters

Scenario for Concrete Mix Design

Based on the design flow in Section 4.1, concrete mix proportions are simulated. The target structures are assumed as underground structures and two types (A and B) are considered. A structure has 75 years and B structure has 100 years for intended service life. Design cover depths are assumed as 50 mm for A structure and 30 mm for B structure. A structure has 65% of R.H. and 12.7°C of temperature. B structure has 75% of R.H. and 22°C of temperature, which are normal exterior conditions in underground site. Extremely high CO_2concentration of 2,700 ppm is assumed for A structure and 980 ppm which is normal condition in underground structure is assumed for B structure. Durability limit state is determined as the condition when carbonation proceeds to steel location [23, 27]. For governing equation, mesolevel equation from CEB [23] is adopted as follows:

$$d_C = \sqrt{2k_1k_2k_3\Delta c} \cdot \sqrt{\frac{D_{CO_2}t}{a}\left(\frac{t_0}{t}\right)^n}, \quad (3)$$

where d_C is carbonation depth (mm), k_1 is constant for local condition, k_2 is constant for curing condition, k_3 is constant for locally different w/c ratio, c is CO_2 concentration (kg/m^3), D_{CO_2} is CO_2 diffusion coefficient (m^2/sec), a is carbonation reaction function with hydrate amount, n is constant for cyclic drying and wetting, t_0 is reference time (1 year), and t is exposed period (year).

For considering the effect of temperature on carbonation, a parameter like (4) is considered [4]. Consider

$$f(T) = D_{\text{ref}} \exp\left[\frac{U}{R}\left(\frac{1}{T_{\text{ref}}} - \frac{1}{T}\right)\right], \quad (4)$$

where D_{ref} is referential CO_2 diffusion coefficient, U is activation energy of CO_2(8500 Cal/mol·K), R is universal gas constant, T_{ref} is reference temperature (298 K), and T is exterior temperature (K).

In (3), the target structure is assumed to have normal construction level and to be sheltered from rain and k_1, k_2, k_3, and n can be assumed as 1.0 and 0.0, respectively [23]. Considering the temperature parameter, (3) can be written as follows:

$$d_C = \sqrt{2\Delta c} \cdot \sqrt{\frac{D_{CO_2}f(T)}{a}t}, \quad (5)$$

where a can be expressed as follows (CEB 1997):

$$a = 0.75 \cdot C \cdot \text{CaO} \cdot \alpha_H \frac{M_{CO_2}}{M_{\text{CaO}}}, \quad (6)$$

where C is unit content of cement (kg/m3), CaO is content of CaO(calciumoxide, 0.65), α_H is hydration rate (0.85), and M is molar weight (CO2: 44 g/mol, CaO: 56 g/mol).

In (5), c, a, t, and T are given by design parameter. Considering the durability limit state (carbonation depth = cover depth), intended diffusion coefficient can be calculated.

The design parameters above are summarized in Table 8.

Table 8: Design parameters for carbonation design

(a)		
Type	Structure A	Structure B
Intended service life (year)	75	100
Design cover depth (mm)	50	30
Exterior condition	R.H.: 65%	R.H: 75%
	Temp.: 12.7°C	Temp.: 22°C
CO2 concentration (ppm)	2,700	980
Durability limit state	carbonation depth = cover depth	

(b)				
Assumed cement weight (kg/m3)	300	330	335	370
Intended diffusion coefficient × 10–11 (m2/sec)	1,742	1,916	1,248	1,378

In (5), two unknown variables exist so that unit content of cement is assumed referring to conventional mix proportions in domestic condition [25]. Four different contents of cement are assumed and the related intended diffusion coefficients are derived through (5).

Derivation of Optimum Mix Proportions

In this section, optimum mix proportions are derived through GA technique. The fitness function of (2) is utilized for obtaining mix components with fixed R.H. and cement content.

The results of mix proportions are listed in Table 9.

Table 9: Mixture design through proposed GA technique

Case		Intended diffusion coefficient (10–11 m2/sec)	w/c (%)	Cement (kg/m3)	S/a (%)	Coarse aggregate (kg/m3)	R.H. (%)
A	Structure A	1,742	42.3	300	38.4	1191.0	65
	Input range		42–58	300-300	37–43	800–1,200	65-65
B	Structure A	1,916	52.7	330	38.1	960.2	65
	Input range		42–58	330-330	37–43	800–1,200	65-65
C	Structure B	1,248	49.6	335	41.2	1172.8	75
	Input range		42–58	335-335	37–43	900–1,200	75-75
D	Structure B	1,378	51.4	370	40.1	875.3	75
	Input range		42–58	370-370	37–43	800–1,200	75-75

As shown in Table 9, intended diffusion coefficient and unit cement content are given and mix proportions for concrete can be obtained through GA technique. When this technique is applied, convergence of relative error to 0.0 should be checked.

In this paper, fitness function for CO_2 diffusion coefficient is derived based on the previous test results, and then concrete mix design is proposed through GA technique. However, this technique is only for OPC concrete mix design and has limitation of range for mix proportion. The applicable ranges of unit content of cement and w/c ratio are 277 kg/m^3~425 kg/m^3 and 0.42~0.58 since both the fitness function and the process for generating each mix proportion are governed by test dataset which is previously adopted.

With more data-set containing CO_2 diffusion coefficient and an accurate fitness function, the proposed technique would be much improved. This technique is applied for mix proportion of concrete under carbonation. With similar procedures, this can be applied to generation of mix proportions which can guarantee the service life of RC structures exposed to different deteriorations like chloride attack, freezing and thawing action, and sulfate attack.

CONCLUDING REMARK

The conclusions on concrete mix optimization technique for service life of RC structures under carbonation using genetic algorithm are as follows.

- Based on the previous experimental results, fitness function for CO_2 diffusion coefficient containing the variables like mix proportions (w/c ratio, unit content of cement, sand/aggregate ratio, and unit content of coarse aggregate) and R.H. (relative humidity) is derived. Through consideration of the parameters of R.H., variation of relative errors decreases.
- Through GA technique, three concrete mix proportions are simulated for verification. The simulated results provide below 10.1% of relative errors for each mix component such as w/c ratio, unit content of cement, sand ratio to total aggregate, and unit content of coarse aggregate.
- Assuming the exposure conditions of carbonation and design parameters, intended diffusion coefficients are determined and optimum concrete mix proportions which satisfy intended service life are obtained through GA technique. The results from this study are only applicable to OPC concrete. If data-set with mineral and chemical admixtures is prepared, this technique can be applied more widely to durability design for RC structures under carbonation.

ACKNOWLEDGMENTS

This research was supported by a grant (Code 11-Technology Innovation-F04) from Construction Technology Research Program funded by Ministry of Land, Infrastructure and Transport.

REFERENCES

1. K. Kobayashi and Y. Uno, "Mechanism of carbonation of concrete," Japan Society of Civil Engineers, vol. 1, no. 1, pp. 139–151, 1990.
2. I. Izumi, D. Kita, and H. Maeda, Carbonation, Kibodang Publication, 1986.
3. T. Ishida and K. Maekawa, "Modeling of PH profile in pore water based on mass transport and chemical equilibrium theory," Japan Society of Civil Engineers, vol. 1, no. 37, pp. 151–166, 2001.
4. T. Saeki, H. Ohga, and S. Nagataki, "Change in micro-structure of concrete due to carbonation," Japan Society of Civil Engineers, vol. 1, no. 18, pp. 1–11, 1991.
5. CEB, Durable Concrete Structures-Design Guide, Thomas Telford, London, UK, 2nd edition, 1992.
6. V. G. Papadakis, C. G. Vayenas, and M. N. Fardis, "Physical and chemical characteristics affecting the durability of concrete," ACI Materials Journal, vol. 88, no. 2, pp. 186–196, 1991.
7. T. Ishida and K. Maekawa, "Modeling of durability performance of cementitious materials and structures based on thermo-hygro physics," in Proceedings of the 2nd International RILEM Workshop on Life Prediction and Aging Management of Concrete Structures, 2003.
8. F. P. Glasser, J. Marchand, and E. Samson, "Durability of concrete—degradation phenomena involving detrimental chemical reactions," Cement and Concrete Research, vol. 38, no. 2, pp. 226–246, 2008.
9. S. J. Kwon, S. S. Park, S. H. Nam, and H. J. Cho, "A study on survey of carbonation for sound, cracked, and joint concrete in RC column in metropolitan city," Journal of the Korea Institute

For Structural Maintenance and Inspection, vol. 11, no. 3, pp. 116–122, 2007.

10. H. W. Song, S. J. Kwon, K. J. Byun, and C. K. Park, "Predicting carbonation in early-aged cracked concrete," Cement and Concrete Research, vol. 36, no. 5, pp. 979–989, 2006. · ·
11. O. B. Isgor and A. G. Razaqpur, "Finite element modeling of coupled heat transfer, moisture transport and carbonation processes in concrete structures," Cement and Concrete Composites, vol. 26, no. 1, pp. 57–73, 2004. · ·
12. S. J. Kwon and U. J. Na, "Prediction of durability for RC column with crack and joint under carbonation based on probabilistic approach," International Journal of Concrete Structures and Materials, vol. 5, no. 1, pp. 11–18, 2011.
13. Y. F. Houst and F. H. Wittmann, "Influence of porosity and water content on the diffusivity of CO_2 and O_2 through hydrated cement paste," Cement and Concrete Research, vol. 24, no. 6, pp. 1165–1176, 1994.
14. S. H. Jung, M. K. Lee, and B. H. Oh, "Measurement device and characteristics of diffusion coefficient of carbon dioxide in concrete," ACI Materials Journal, vol. 108, no. 6, pp. 589–595, 2011.
15. S. H. Jung, Diffusivity of carbon dioxide and carbonation in concrete through development of gas diffusion measuring system [Ph.D. thesis], Deptartment of Civil Engineering, Seoul National University, Seoul, Republic of Korea, 2002.
16. S. J. Kwon and H. W. Song, "Analysis of carbonation behavior in concrete using neural network algorithm and carbonation modeling," Cement and Concrete Research, vol. 40, no. 1, pp. 119–127, 2010. · ·
17. I. C. Yeh, "Computer-aided design for optimum concrete mixtures," Cement and Concrete Composites, vol. 29, no. 3, pp. 193–202, 2007. · ·
18. C. H. Lim, Y. S. Yoon, and J. H. Kim, "Genetic algorithm in mix proportioning of high-performance concrete," Cement and Concrete Research, vol. 34, no. 3, pp. 409–420, 2004. · ·
19. S. J. Kwon and S. C. Kim, "Concrete mix design for service life of RC structures exposed to chloride attack," Computers and Concrete, vol. 10, no. 1, pp. 687–607, 2012.

20. E. Cantú-Paz and D. E. Goldberg, "Efficient parallel genetic algorithms: theory and practice," Computer Methods in Applied Mechanics and Engineering, vol. 186, no. 2–4, pp. 221–238, 2000.
21. D. E. Goldberg, Genetic Algorithms in Search, Optimization and Machine Learning, Addison-Welsley, Reading, Mass, USA, 1989.
22. Y. Abe, "Result of reference review on crack width effect to carbonation of concrete," in Proceedings of the Symposium on Rehabilitation of Concrete Structures, 1999.
23. CEB Task Group 5.1, 5.2, New Approach to Durability Design, CEB, Sprint-Druck, Stuttgart, Germany, pp. 53–62, May 1997.
24. KICT-Korea Institute of Construction Technology, "Durability improvement of concrete with sea/normal sand," Tech. Rep. R&D/97-0001, section 1–3, Bon Press, Seoul, Republic of Korea, 1997.
25. KSCE-Korea Society of Civil Engineering, "Study on durability improvement for underground structure," Tech. Rep., Seoul Metropolitan Government Office of Subway Construction, KSCE Press, Seoul, Republic of Korea, 1999.
26. KCI-Korea Concrete Institute, Concrete Specification-Durability, 2004.
27. RILEM, "Durability design of concrete structures," Report of RILEM Technical Committee 130-CSL, E&FN, 1994, pp. 75–78.
28. JSCE-Concrete Committee, Standard Specification for Concrete Structures, 2002.
29. Mathworks, Genetic Algorithm and Direct Search Toolbox 2, User›s Guide, 2007.

Chapter 3

Optimization Design and Application of Underground Reinforced Concrete Bifurcation Pipe

Chao Su[1], Zhenxue Zhu[1], Yangyang Zhang[1], and Niantang Jiang[2]

[1]College of Water Conservancy and Hydropower Engineering, Hohai University, No. 1 Xikang Road, Nanjing 210098, China

[2]Shanghai Investigation, Design & Research Institute, No. 388 Yixian Road, Shanghai 200434, China

ABSTRACT

Underground reinforced concrete bifurcation pipe is an important part of conveyance structure. During construction, the workload of excavation and concrete pouring can be significantly decreased according to optimized pipe structure, and the engineering quality can be improved. This paper presents an optimization mathematical model of underground reinforced concrete bifurcation pipe structure according to real working status of several common pipe structures from real cases. Then, an optimization design system was developed based on Particle Swarm Optimization algorithm. Furthermore, take the bifurcation pipe of one hydropower station as an example: optimization analysis was conducted, and accuracy and stability of the optimization design system were verified successfully.

INTRODUCTION

With the development of large hydropower stations especially pumped storage power stations, more and more bifurcation pipes are constructed. In conveyance system of hydropower stations, bifurcation pipe usually locates on the end of pressure pipe and is near the powerhouse, so it undertakes high internal and external water pressure. Bifurcation pipe is a complex space structure and complicated stresses are caused when it suffers from different water conditions. There are many constraints when it is designed, and requirements of hydraulic power and structure are needed to be satisfied; however, there is certain contradiction between these two aspects; they cannot be met mutually. In 1990, Zagars et al. [1] analyzed reinforced concrete bifurcation pipe of Guxu storage power station using structural mechanical method. In 1996, linear and nonlinear finite element methods were used for evaluating stress status of surrounding rock, lining, and reinforcing steel bars of bifurcation pipe under internal and external effect by Lu and Liu [2]. In 2001, Xiao [3] developed the optimization analysis method used for assessing the best structure of large-scale underground reinforced concrete conduit based on failure conditions of reinforced concrete and elastoplastic principle of rock energy dissipation. In 2006, Haixia and Zhefei [4] established an optimization design model of crescent rib branches for steel pipe and developed a corresponding program. In 2009, Ren et al. [5] analyzed stress and deformation of surrounding rock and reinforcement of lining material, stress, and crack in reinforced concrete bifurcation structure of Pushihe pumped storage power station.

In bifurcation pipe design, design safety and reasonable structure based on real structure layout are the main purposes of most designers. Bifurcation pipe optimization can reduce not only structure volume, but also cavern excavation volume [6]. However, most of the current optimization designs are traditional; they only make simple comparisons of some schemes without mathematical programming method and material properties cannot be fully used. A reasonable and efficient method is needed and will be significant for optimization design of bifurcation pipe.

PARAMETERIZED FEM MODEL AND MESH PROGRAM

Shape-control parameters should be defined according to different bifurcation pipe structures before establishing model, and then the coordinates of shape-control points can be calculated by shape-control parameters. Three-dimensional model containing reinforced concrete blocks of bifurcation pipe is built and an automatic mesh program is developed [7–9].

Basic design parameters of bifurcation pipe usually are main inlet pipe diameter R_1, branch pipe outlet diameter R_2R_3, bifurcation angle b, joint tangent sphere radius$R$,main half cone angle $a1$, half cone angle of branch cone a_2a_3, waist turning angle $\omega_2\omega_3$, wall thickness $t_1t_2t_3$ (not shown in Figure 1), and so forth (see Figure 1). Shape-control point sand geometric sizes can be calculated according to geometry relations. Shape-control point calculation of bifurcation pipe is done. Calculation of some intermediate variables and auxiliary geometric sizes is necessary [10].

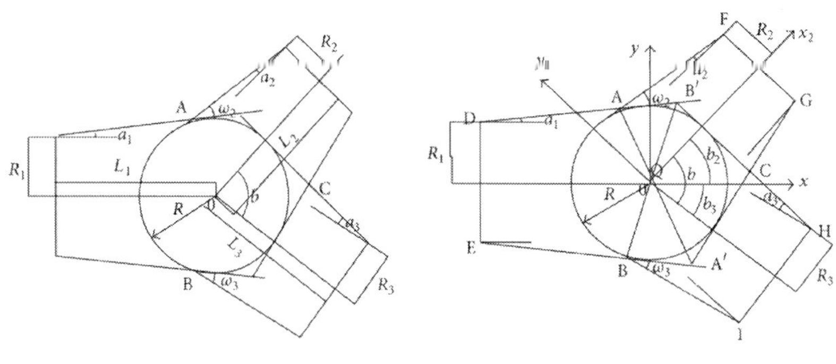

Figure 1: Diagrams of Y-type bifurcation pipe and shape-control points.

Control point coordinates calculation program was written by Visual Basic. Bifurcation pipe control points and surrounding rock coordinates were input through visualization parameter input interface and data files such as CGFK.TXT, KZD.TXT were exported. According to the defined mesh scheme, the whole three-dimensional finite element mesh was obtained using automatic mesh program CGWG. EXE written by FORTRAN language, converting the mesh data into INP format files of ABAQUS and then the whole mesh can be visualized in ABAQUS [11–13].

OPTIMIZATION DESIGN PROGRAM

The process of optimization design includes randomization of initial parameters, calculation of structure, judgment of constraints, searching better optimization design scheme, and obtaining the best optimization design results. Implementation usually goes through the following steps: (1) establish a reasonable mathematical model and transform real engineering problems into mathematical problems; (2) solve mathematical problems by certain reasonable and effective optimization method; (3) develop a program using structural analysis method and implement optimization design calculation.

Mathematical Model

Optimization design in this paper was based on a defined topology structure, so design variables should be chosen according to the topology. Bifurcation pipe shape-control parameters were chosen as design variables. The least concrete volume was set as objective function in order to make full use of reinforced concrete bifurcation pipe material properties.

Constraints contained geometric and strength constraints.Geometric constraints were as follows: (1) half coneangle is $\theta_1 < a_i < \theta_2$ and (2) bifurcation angle is $\theta_3 < b < \theta_4$;(3) the shortest bus bar of pipes was no less than 300mm;and (4) wall thickness of concrete bifurcation pip is $d_1 < t_i < d_2$ (θ_1, θ_2 were the range of half cone angle. θ_3, θ_4 werethe range of bifurcation angle. $d1$, $d2$ were the range of wallthickness). Stress constraints included the requirements thattensile stress should be lower than concrete ultimate tensilestrength design value and compressive stress should be lowerthan concrete ultimate compressive strength design value

Optimization Design Algorithm

Mathematical model of bifurcation pipe optimization is often nonlinear programming problems. Applicable optimization algorithm includes generalized reduced gradient method, complex method, Particle Swarm Optimization, penalty method, sequential linear programming, and sequential quadratic programming. Particle Swarm Optimization algorithm is adopted in this paper. Particle Swarm Optimization has the advantages of parallel processing, good stability, and the most probability of finding global optimal solution [14].

A group of random particles was firstly initialized in PSO algorithm and particle's searching following optimal particle was conducted in the solution space. In searching process, every particle updated its position and velocity according to the current optimal particle position. The global optimal solution was obtained by continuous iteration and searching [15].

The optimization process of N particles group in D-dimension searching space was taken as an example todescribe the process of optimization algorithms. The state ofparticle i in time t was set as follows:

velocity: $v_i^k = (v_{i1}^k, V\ v_{i2}^k, \ldots, v_{ij}^k, \ldots, v_{iD}^k)$

position: $x_i^k = (x_{i1}^k, x_{i2}^k, \ldots, x_{ij}^k, \ldots, x_{iD}^k)$

personal best position: $pbest\,x_i^k = (pbest\,x_{i1}^k, pbest\,x_{i2}^k, \ldots, pbest\,x_{ij}^k, \ldots, pbest\,x_{iD}^k)$

global best position: $gbest\,x_i^k = (gbest\,xk_1^i, gbest\,x_{i2}^k, \ldots, gbest\,x_{ij}^k, \ldots, gbest\,x_D^k)$,

where $i = 1, 2, \ldots, N$; $j = 1, 2, \ldots, D$; $k = 1, 2, \ldots, it\ max$;$it\ max$ is the maximum number of iterations, i represents the ith particle, j represents velocity or the jth dimension, and $k$represents the count of iterations.

In every iteration process, how to update the position of particle is the key factor affecting optimization algorithm effectiveness. Every particle updates its position according to two optimal solutions (global best position ***gbest*** and personal best position ***pbest***). Updated velocity and new position can be defined by following formula:

$$v_{ij}^{k+1} = v_{ij}^{k} + c_1 * r_1 \left(pbestx_{ij}^{k} - x_{ij}^{k}\right)$$
$$+ c_2 * r_2 * \left(gbestx_{j}^{k} - x_{ij}^{k}\right),$$
$$x_{ij}^{k+1} = x_{ij}^{k} + v_{ij}^{k+1}. \tag{1}$$

In order to coordinate the global and local optimization capability of PSO algorithm, inertia weight coefficient ω is set to determine the amount of current velocity particles inherit. Transform the velocity equation to formula (2) and keep the position updating equation:

$$v_{ij}^{k+1} = \omega_k v_{ij}^{k} + c_1 * r_1 \left(pbest_{ij}^{k} - x_{ij}^{k}\right)$$
$$+ c_2 * r_2 * \left(gbest_{j}^{k} - x_{ij}^{k}\right). \tag{2}$$

In the equation, v_{ij}^{k} and x_{ij}^{k} are the *j*th dimension velocityand position of particle *i* in the *k*th iteration respectively. $c_1$and c_2 are constant, usually $c_1c_2 \in [0, 4]$; r_1 and r_2 are randomnumber between 0 and 1; ***pbest***x_{ij}^{k} is the personal best positionof particle P_i in the *j*th dimension; ***gbest***x_{ij}^{k} is the global bestposition of the group.

Linear decreasing inertia weight is usually adopted in PSO iteration calculation process at present. It can accelerate the convergence of iterative calculation and get global best solution quickly. Y. Shi has suggested that inertia weight ω decreases linearly from 0.9 to 0.4 for better algorithm performance. The linear decreasing formula is as follows:

$$\omega_k = \omega_{\max} - \frac{\omega_{\max} - \omega_{\min}}{it\max} * k. \tag{3}$$

The position updating can be described as in Figure 2.

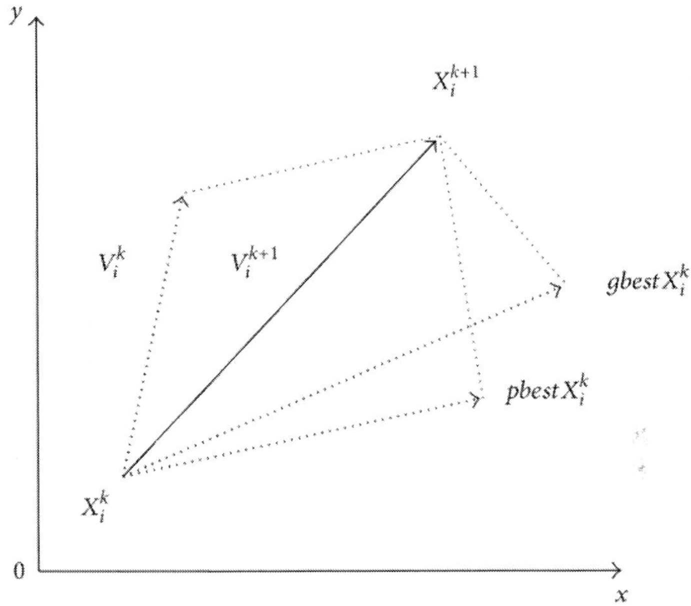

Figure 2: Particle updating schematic diagram.

Reinitialization is carried out on particles which is out of range or do not satisfy constraints during development of the program. To ensure the minimum objective function value, convergence criterion is that the difference between global optimal solutions of last generation and next generation should be less than calculation tolerance δ. At the same time, to avoid too many iterations, the maximum count of iteration is 1000 in program.

Structural Analysis Method Selection

FEM was used in structure optimization design and calculation in this paper [16, 17]. FEM is not only with complete and reliable theoretical basis, but also with simple and standardized basis. It can meet the accuracy requirements of engineering calculation and its convergence can be guaranteed. Calculating results can show real structure stress and deformation characteristics.

ENGINEERING APPLICATIONS

Project Overview

A hydropower station with total installed capacity of 1200 MW, unit capacity of 300 MW has four units. The depth of #1 bifurcated pipe is approximately 414 m–418 m and #2 is approximately 415 m. The impact of complex geological conditions can be considered in the finite element method for structural calculation.

C30 concrete is used for lining in bifurcation pipes construction. Lining thickness is 0.5m $\leq t \leq$ 1m. Two 20mmhoop reinforcement scheme is adopted to prevent concretecracking under high water pressure during operation period.Themain mechanical properties of C30 concrete and steel areshown in Table 1.

Table 1: Main mechanical properties of C30 concrete and rebar

Mechanical index	C30 concrete	Rebar
Density	2.5 g/cm^3	—
Elastic modulus E_0	28 GPa	210 GPa
Poisson's ratio μ	0.167	0.3
Tensile strength	1.5 MPa	210 MPa
Compressive strength	15 MPa	210 MPa

The optimization design calculation in this paper focused on normal operation cases. The influence of gravity of rock was considered and it was assumed that all gaps between bifurcation pipe and rock were ignored. Linear calculation was implemented for bifurcation pipe under internal water head around 520 m (including water hammer pressure) and gravity of surrounding rock without consideration of external water pressure. Finally, structural stress and strain were analyzed.

Mathematical Model of Bifurcation Pipe

According tothe initial design scheme, parameters of bifurcation pipesize model were confirmed. Structure type is shown inFigure 3.

Parameters which can be used as optimization design parameters are half angle of main cone and branchcone a_1a_2, bifurcation angle b, main cone wall thickness t_1, and branch cone wall thickness t_2t_3 (see Figure 4).

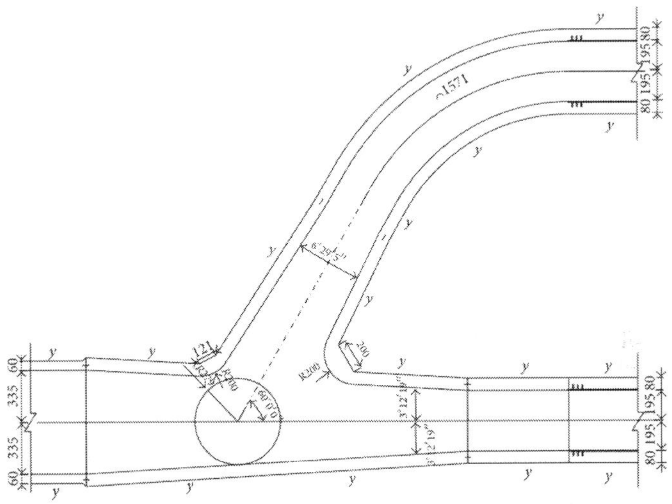

Figure 3: Plane design drawing of bifurcation pipe.

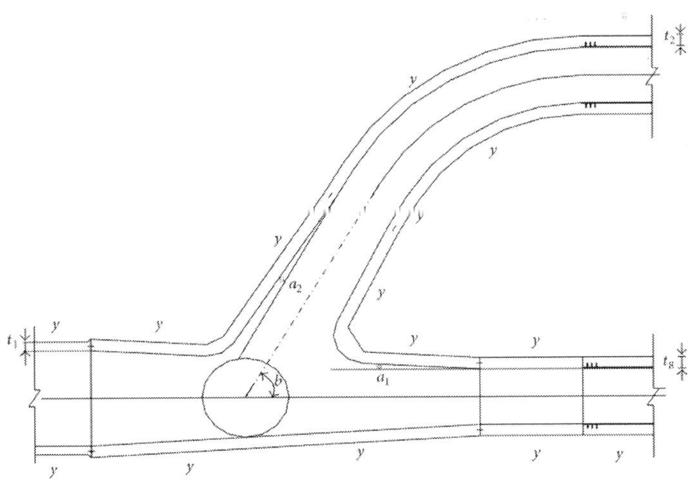

Figure 4: Schematic diagram of optimization design parameters.

Bifurcation pipe concrete volume was set as objective function. According to bifurcation pipe design codes, the range of design variables is shown in Table 2.

Table 2: Value range of design variables

Optimization variables	**Range**
Bifurcation angle b	50~70°
Half angle of main cone a_1	2~5°
Half angle of branch cone a_2	2~5°
Main cone wall thickness t_1	0.5~1 m
Branch cone wall thickness t_2	0.5~1 m
Branch cone wall thickness t_3	0.5~1 m

C30 concrete lining is used in this project, so stress control standards in this optimization design are C30 concrete ultimate tensile strength and compressive strength. However, because of reinforcement, tensile stress control standards were calculated using equivalent stress method [18]. According to the amount of bifurcation pipe reinforcement, corresponding design value of ultimate concrete tensile stress can be increased to 7 MPa. Considering that reinforcement is mainly subject to tensile stress and its contribution to compressive resistance could be ignored, the corresponding concrete ultimate compressive stress is unchanged. Under these stress constraints, bifurcation pipe size should meet the specification requirements by traditional calculating method. Bifurcation pipe optimization design implementation flow is shown in Figure 5.

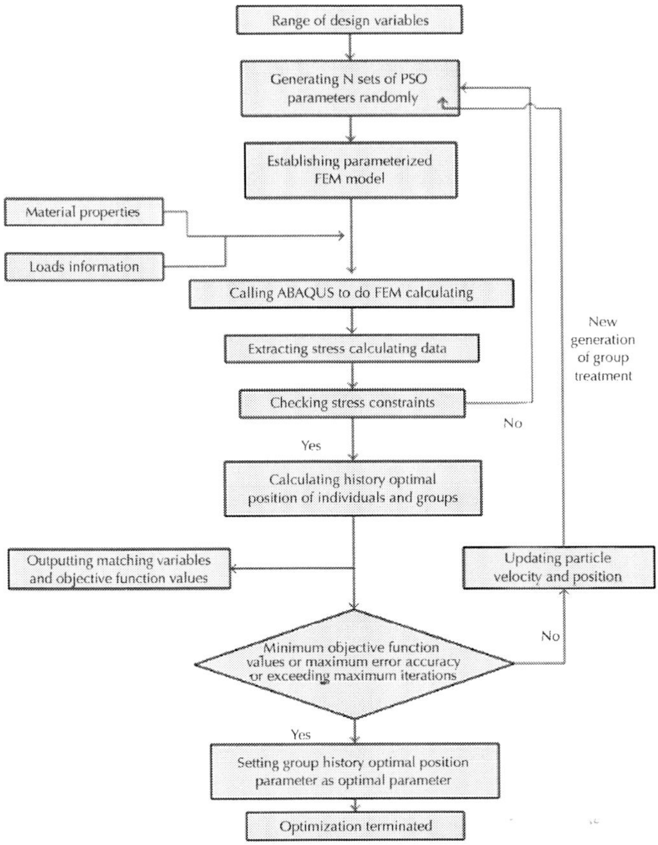

Figure 5: Flowchart of optimum design program.

Optimization Results Analysis

Bifurcation pipe of a hydropower station was analyzed using the optimization design system and the better design parameters were obtained in Table 3 after 178 iterations. Bifurcation angle is significantly reduced, walls are much thinner, and the amount of concrete is also reduced. The amount of concrete is 2677.41 m^3 before optimization and it becomes 1833.30 m^3 after optimization, so the amount of concrete reduces to 43.2% after shape optimization. Reinforcement checking calculation was conducted by using design parameters after optimization.

Table 3: Comparison of initial shape and optimization shape.

Variable	Before optimization	After optimization
Bifurcation angle b (°)	60	47.86
Half angle of main cone a_1 (°)	3.2	4.65
Half angle of branch cone a_2 (°)	3.2	4.57
Main cone wall thickness t_1 (m)	0.6	0.57
Branch cone wall thickness t_2 (m)	0.8	0.63
Branch cone wall thickness t_3 (m)	0.8	0.63

Surrounding Rock

The distribution of surrounding rock stress and strain is similar before and after optimization with some differences in extreme values. After excavation, the whole rock is almost in compressive status, only a small range of tensile status occurs in the bottom of the cavern excavation area at bifurcation area. The maximum tensile stress before and after optimization is 0.48 MPa and 0.59 MPa. Compressive stress concentration occurs in the bifurcation area and the maximum compressive stress reaches −43.71 MPa, −57.56 MPa. Large displacement is caused in top and bottom of cavern toward to the cave. Extreme displacement occurs at the top of the bifurcation of the cavern, and displacement before and after optimization is 2.31 mm and 2.12 mm. After parameters optimization, walls are significantly thinner, cavern excavation volume decreases, and cavern excavation radius is smaller. The top and bottom cavern displacement after excavation has been reduced. Cavern excavation radius has a certain relationship with the surrounding rock displacement.

Bifurcation Pipe

The distribution of bifurcation pipe stress and strain is almost unchanged before and after optimization. Two layers of 20 mm hoop reinforcement were built using T3D2 element in modeling and were inserted into inner side of bifurcation pipe. Analysis of bifurcation pipe with hoop reinforcement is as follows. There is little change in the whole stress distribution. Whole bifurcation stress after inserting hoop reinforcement is significantly reduced. The vast majority of the tensile stress values is 2.0 MPa or less, but there is still stress concentration phenomenon at the bifurcation area and maximum stress here can be eliminated through suitable reinforcement direction arrangement. Extreme displacement is 0.535 mm at intersecting line in top and bottom of inner surface. All rebars are subject to tensile stress and maximum stress reaches 7.52 MPa (see Figures 6 and 7).

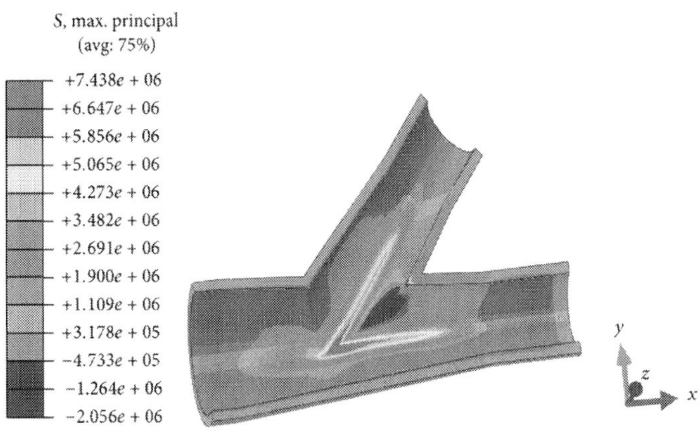

Figure 6: Principal tensile stress of inside surface.

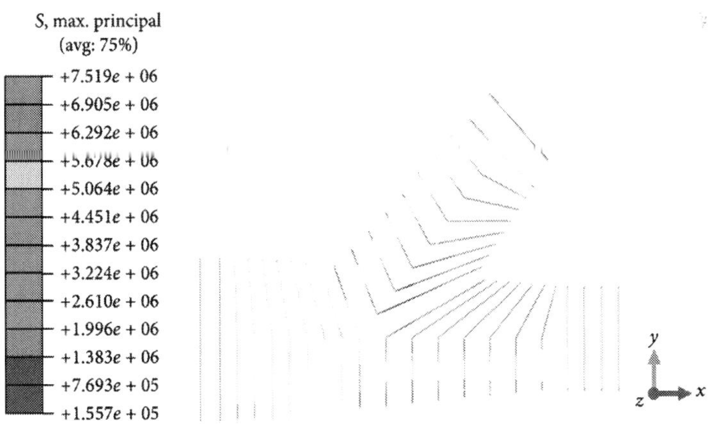

Figure 7: Principal tensile stress of hoop rebar.

For the inevitable stress concentration in bifurcation and waist areas, some general engineering measures can be taken such as sharp corners rounded, haunch crotch, outsourcing concrete in bifurcation point, configuring hoop, and axial steels. Reducing the external water pressure is also an important measure for stabilizing bifurcation pipe.

CONCLUSIONS

Bifurcation pipe is commonly used in water conveyance system of hydropower station and pumped storage power plant. Its structural safety and stability are directly related to hydropower security and stability, so safety and stability of bifurcation pipe design are very important. Three-dimensional FEM parametric reinforced concrete bifurcation pipe program is established in this paper. Particle Swarm Optimization method is selected as the bifurcation pipe structure optimization algorithm and corresponding optimization program is written. FEM calculation is done by calling ABAQUS/Standard module through interface technology between optimization program and ABAQUS to achieve the optimization goals. In the verified project example, the amount of concrete of optimized bifurcation pipe was significantly reduced. At the same time, the comparison of stress and displacement results obtained before and after optimization proved that the program is feasible and stable.

ACKNOWLEDGMENTS

The present work is supported by the National Natural Science Foundation of China (no. 51279053) and the Fundamental Research Funds for the Central Universities (no. 2014B36514).

REFERENCES

1. A. Zagars, C. H. Yeh, and H. Lu, "Design of high pressure reinforced concrete bifurcation pipe," Water Power, no. 10, pp. 13–18, 1990.

2. X. Lu and Q. Liu, "Stress analysis for underground reinforced concrete partition manifold," Journal of Hohai University (Natural Sciences), vol. 9, pp. 24–29, 1996.
3. M. Xiao, "Optimization analysis of large-scale underground reinforced concrete branching conduit,"Journal of Hydraulic Engineering, vol. 32, no. 12, pp. 8–14, 2001.
4. Y. Haixia and L. Zhefei, "Optimum design of crescent rib branches for steel pipe," Chinese Journal of Solid Mechanics, vol. Z1, pp. 180–180, 2006.
5. X.-H. Ren, H.-B. Xu, and J.-Q. Shu, "Numerical analysis on surrounding rock and reinforced concrete bifurcation structure of underground tunnels," Journal of Sichuan University (Engineering Science), vol. 41, no. 4, pp. 8–13, 2009.
6. Q. Liu, Hydropower Plant, China Water Power Press, 2004.
7. B. Zhang, H. Chen, and L. Zou, "Simple implementation of complex hydraulic structure finite element discretization," Water Resources and Hydropower Engineering, vol. 29, no. 12, pp. 4–6, 1998.
8. Z. Yufeng and Z. Yiwen, "Review for typical methods of finite element mesh automatic generation and research topics in future," Engineering Journal of Wuhan University, vol. 38, no. 2, pp. 54–59, 2005.
9. X. Peng and X. Qian, "Zoning-direct method of automatic FE mesh generation," Engineering Mechanics, no. 4, pp. 1396–148, 1993.
10. W. Hegao, Y. Xingyi, and W. Shougen, "Research and application on system for computer aided design of rib reinforced branch pipe," in Proceedings of the The 6th National Hydropower Penstock Papers, p. 6, 2006.
11. Z. Bofang, Principles and Applications of Finite Element Method, China WaterPower Press, 1998.
12. E. Nie and C. Su, "Application of VB and TrueGRID to 3D finite grid automatic mesh generation for lock head," Port & Waterway Engineering, no. 10, pp. 138–142, 2009.
13. S. Chao, N. Jiang, and J. Wang, "Optimization design of chamfer triangular gate lock structure," Port & Waterway Engineering, no. 5, pp. 124–129, 2013.

14. Q. Feng and Y. Li, "Research on the application of IPSO in engineering optimization problem," Chinese Journal of Scientific Instrument, vol. 26, no. 9, pp. 984–990, 2005.
15. G. Xu, J.-P. Qu, and Z.-T. Yang, "Improved adaptive particle swarm optimization algorithm," Journal of South China University of Technology (Natural Science), vol. 36, no. 9, pp. 6–10, 2008. View at Scopus
16. C. Peijiang, B. Tengfeng, and Y. Peng, "3D finite element analysis of underground embedded steel bifurcation pipe," Hydropower Automation and Dam Monitoring, vol. 3, pp. 64–66, 74, 2006.
17. Q. Yan, "3D finite element analysis of underground embedded steel bifurcation structure," Water Power, vol. 6, pp. 29–32, 2003.
18. F. Zuoxin and Q. Xiangdong, "Application of finite element method to the design of arch dams," Journal of Hohai University (Natural Sciences), vol. 19, no. 2, pp. 8–15, 1991.

Chapter 4

Methodology for Assessing the Probability of Corrosion in Concrete Structures on the Basis of Half-Cell Potential and Concrete Resistivity Measurements

Lukasz Sadowski

Institute of Building Engineering, Wroclaw University of Technology, Plac Grunwaldzki 11, 50-377 Wroclaw, Poland

ABSTRACT

In recent years, the corrosion of steel reinforcement has become a major problem in the construction industry. Therefore, much attention has been given to developing methods of predicting the service life of reinforced concrete structures. The progress of corrosion cannot be visually assessed until a crack or a delamination appears. The corrosion process can be tracked using several electrochemical techniques. Most commonly the half-cell potential measurement technique is used for this purpose. However, it is generally accepted that it should be supplemented with other techniques. Hence, a methodology for assessing the probability of corrosion in concrete slabs by means of a combination of two methods, that is, the half-cell potential method and the concrete resistivity method, is proposed. An assessment of the probability of corrosion in reinforced concrete structures carried out using the proposed methodology is presented. 200 mm thick 750 mm × 750 mm reinforced concrete slab specimens were investigated. Potential E_{corr} and concrete resistivity in each point of the applied grid were measured. The experimental results indicate that the proposed methodology can be successfully used to assess the probability of corrosion in concrete structures.

INTRODUCTION

Corrosion of the steel reinforcement in concrete is a crucial problem for the construction industry since it poses the most serious risk to the structural integrity of reinforced concrete structures. Inspection and monitoring techniques are needed to assess the corrosion of the reinforcement in order to maintain, protect, and repair buildings and bridge decks so that they remain safe [1, 2]. In the last few years much attention has been given to developing techniques for predicting the remaining service life of concrete structures [3]. Most of the reported research in this area focuses on the corrosion of concrete reinforcement [4].

Several electrochemical techniques for monitoring and assessing the corrosion of steel in concrete structures were presented in [5–7]. The most popular method of in situ corrosion testing is the half-cell potential measurement, the idea of which is illustrated in Figure 1(a). The use of this method and the interpretation of its results are described in ASTM C876 [8]. Such corrosion potential measurements, however, should be supplemented with other nondestructive testing methods [9–12]. The use of half-cell potential measurements for determining the probability of corrosion in concrete was extensively described by Flis et al. [13], Grantham et al. [14], and Živica [15]. The latter also presented dependences between the rate of reinforcement corrosion and temperature. The effectiveness of the test was studied in [16]. It should be noted that half-cell potential values merely provide information about the probability of corrosion and not about the rate of corrosion.

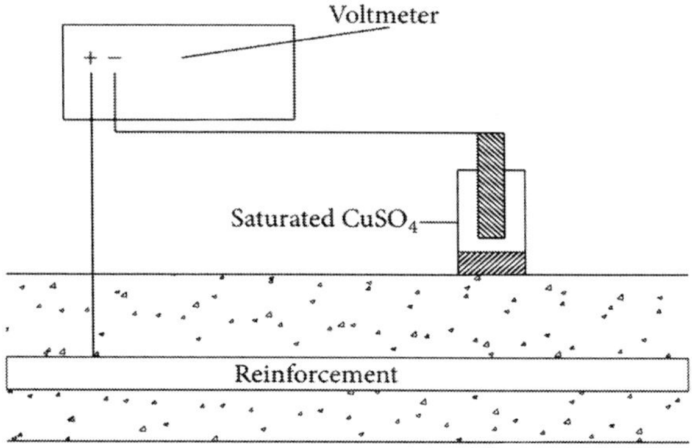

(a)

(a)

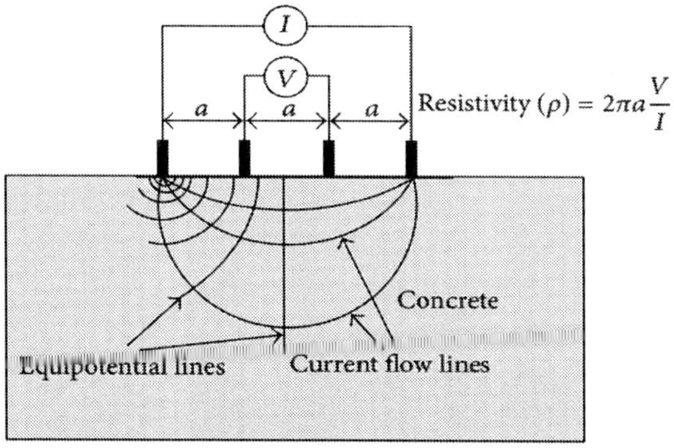

(b)

(b)

Figure 1: Existing corrosion methods: (a) half-cell potential measurement [27] and (b) concrete resistivity measurement [28].

It is well known that the probability of corrosion in concrete structures depends on the ionic conductivity of the concrete electrolyte, the humidity, the temperature, and the quality of the concrete cover. The ionic conductivity is measured quantitatively as the resistivity of the concrete [17]. Concrete resistivity ranges widely from 10^1 to $10^6\,\Omega\,m$, depending on mainly the moisture content [18] and the material of the concrete [19, 20]. One of the promising techniques of measuring concrete resistivity is shown in Figure 1(b). As shown by Feliu et al. [21], concrete resistivity is inversely proportional to the corrosion rate. This was confirmed by Glass et al. [22] who showed that the effect of mortar resistivity is strongly dependent on the relative humidity of the environment, while López et al. [23] showed that the amount of pores in concrete determines its resistivity and corrosion rate. The four-point resistivity method enables one to determine the severity of corrosion in a quick and nondestructive manner. Morris et al. [24] found that rebars undergo active corrosion when concrete resistivity is below 10 kΩcm, whereas at concrete resistivity above 30 kΩcm the probability of their corrosion is low. Extensive research on the resistivity technique, covering experimental analyses [25] and an analysis of the effects of geometry and material properties [26], was done by Zhang et al.

It should be noted that in the literature there are only a few papers in which the probability of corrosion is determined using both concrete resistivity measurements and half-cell potential mapping. One of them is the paper by Millard and Sadowski [29] in which in order to determine the degree of corrosion of the reinforcement, the resistivity of the concrete is measured by the electrodes used in the half-cell potential method. The combined use of half-cell potential and resistivity measurements was presented in [27]. In [28] it was shown that the combination of the method described in [29] and the conventional method of measuring concrete resistivity could be a reliable tool for directly determining the corrosion rate of the reinforcement in concrete. In paper [30] for this purpose Vedalakshmi et al. used the galvanostatic pulse technique with electrochemical impedance spectroscopy and the weight-loss method. Rhazi [31] measured concrete resistivity in the same locations where half-cell potentials were measured, but it should be noted that the measurements were carried out on a concrete bridge deck covered with asphalt.

Considering that in the literature it is hard to find a systematic methodology for assessing the probability of corrosion in concrete slabs through combined half-cell potential and concrete resistivity measurements, this paper proposes such a methodology based on the combined use of the four-point Wenner concrete resistivity method and the half-cell potential method.

METHODOLOGY FOR ASSESSING PROBABILITY OF CORROSION IN CONCRETE SLABS

The proposed methodology for assessing the probability of corrosion in concrete slabs through a combination of two methods, that is, the half-cell potential method and the concrete resistivity method, is illustrated graphically in Figure 2.

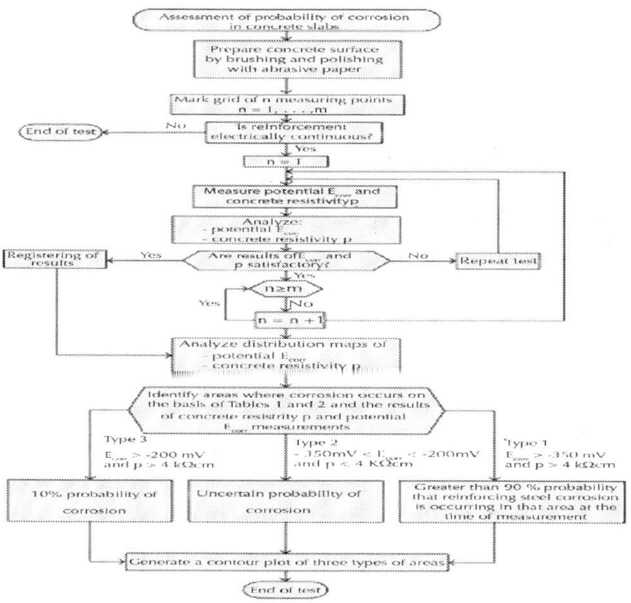

Figure 2: Assessment of probability of corrosion in concrete slabs through half-cell potential and concrete resistivity measurements.

Before measurements, the surface of the concrete slab should be prepared by brushing and polishing with abrasive paper, and a grid of n measuring points spaced at every 75 mm should be marked on the slab surface. Then the electrical continuity of the reinforcement is checked in three randomly selected grid points. Subsequently potential E_{corr} and concrete resistivity are measured. If the results of the E_{corr} and measurements are satisfactory, they are processed using the specialized software and maps of the distribution of the parameter values on the slab surface are produced. The distribution maps of potential E_{corr} and concrete resistivity should be examined. Table 2 summarizes the typical interpretation of half-cell potential readings [8]. The dependences between reinforcement corrosion probability and concrete resistivity measured by the four-point Wenner method are shown in accordance with [32] in Table 3. If there is a probability of corrosion, one should identify the areas in which corrosion may occur. On the basis of Tables 1 and 2 and the results of the concrete resistivity and potential E_{corr} measurements, three areas of different types can be generated:

- area type 1—low values of both parameters, greater than 90% probability that reinforcing steel corrosion is occurring in that area at the time of measurement;
- area type 2—low values of concrete resistivity ρ and high values of measured potential E_{corr}, an uncertain probability of corrosion;
- area type 3—high values of both parameters, a 10% probability of corrosion.

Table 1: Dependence between potential and corrosion probability [8]

Potential E_{corr}	**Probability of corrosion**
E_{corr} <−350 mV	Greater than 90% probability that reinforcing steel corrosion is occurring in that area at the time of measurement
−350 mV≤ E_{corr} ≤−200 mV	Corrosion activity of the reinforcing steel in that area is uncertain

E_{corr} >−200 mV	90% probability that no reinforcing steel corrosion is occurring in that area at the time of measurement (10% risk of corrosion)

Table 2: Dependence between concrete resistivity and corrosion probability [34]

Concrete resistivity , k Ω cm	**Probability of corrosion**
< 5	Very high
5< ρ <10	High
10< <20	Low to moderate
>20	Low

Table 3: Potential E_{corr}

Nr	**Potential E_{corr}, mV**								
	A	B	C	D	E	F	G	H	I
1	-192	-168	-154	-151	-144	-156	-178	-192	-288
2	-240	-167	-169	-167	-168	-187	-201	-216	-276
3	-234	-168	-198	-189	-178	-189	-200	-215	-281
4	-201	-178	-201	-190	-189	-191	-200	-199	-291
5	-168	-165	-181	-191	-192	-189	-194	-192	-360
6	-231	-174	-182	-204	-201	-199	-194	-198	-365
7	-203	-164	-98	-219	-232	-231	-199	-196	-369
8	-216	-167	-197	-207	-216	-245	-234	-192	-384
9	-192	-168	-201	-231	-240	-245	-219	-216	-432

Finally, a contour plot of three types of areas (type 1, type 2, and type 3) should be generated.

The corrosion probability assessment can be practically verified through test pits made in selected places and visual inspection [33].

EXEMPLARY APPLICATION OF THE PROPOSED METHODOLOGY

Materials and Methods

An exemplary application of the proposed methodology is presented below. 200 mm thick 750 mm × 750 mm concrete slab specimen was investigated (Figure 3). The concrete had been designed to strengthen class C 20/25. Portland cement CEM I 42.5R, well-graded sand, and crushed blue granite with a maximum total grading of 5 mm, consistency S3, and w/c = 0.5 had been used to cast the slab specimens. A reinforcement mesh made of A-III 34GS steel rebars 10 mm in diameter spaced at every 150 mm with a 50 mm cover had been embedded inside each slab. The concrete slab specimen was placed in a normal atmosphere and was subjected to a two-hour spray wetting with a sodium chloride solution cycle followed by twenty hours drying cycle to generate corroding area presented in Figure 3.

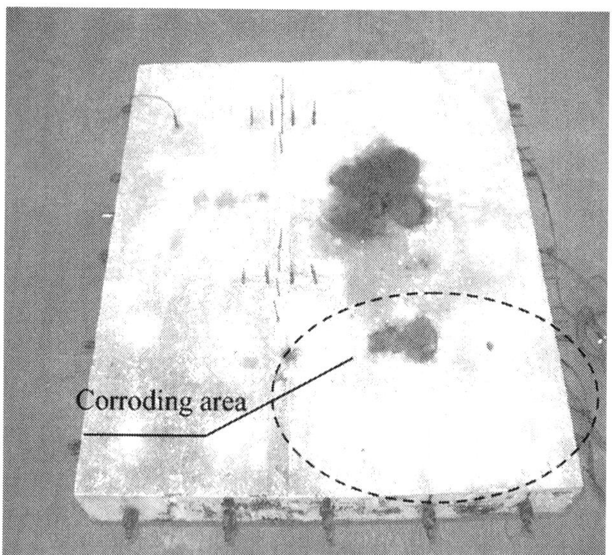

Figure 3: General view of concrete slab specimen.

Measurements were carried out after 90 days of concrete curing, except for compressive strength tests which were done after 28 days. The concrete slabs cured at an air temperature of +18°C (±3°C) and a relative air humidity of 60%. It is important to measure concrete resistivity in constant temperature-humidity conditions since, as shown in [34], with each degree Celsius relative humidity increases by 3% [35]. After the concrete slabs were labelled, a 750 mm × 750 mm test area was demarcated on each of them and a grid of points spaced at every 75 mm was marked on each of the slabs. The columns were denoted with letters from A to I, and the rows were numbered from 1 to 9. A total of 81 measuring points were marked on the surface.

Prior to half-cell potential measurements, the electrical continuity of the reinforcement was checked at three randomly selected grid points. The potential differences in the three points were measured with a digital voltmeter, and all the measured values were found to be below 1 mV. The copper/copper sulphate electrode used in this test is shown in Figure 4(a), and the digital voltmeter with high input impedance is shown in Figure 4(b). Before the measurements, the concrete surface was prepared by brushing and polishing with abrasive paper. The area where each measurement was to be taken was wetted with tap water to ensure better electrical contact. Then the reference electrode and the reinforcement mesh were connected to the high-impedance voltmeter, and the reference electrode was placed on the surface of the concrete.

(a)

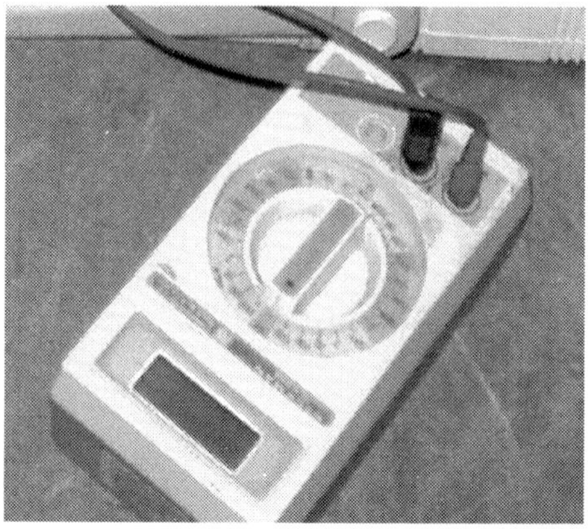

(b)

Figure 4: Half-cell potential measurements: (a) copper/copper sulphate electrode and (b) digital voltmeter.

The concrete resistivity measurements were carried out at frequencies in a range of 50–1000 Hz. Before the measurements, the equipment was calibrated using a 1 kΩ calibration bar. A single deviation was not larger than 2% and the average deviation amounted to less than 1%. Concrete resistivity was measured in each point of the grid (Figure 5).

(a)

(b)

Figure 5: Concrete resistivity measurement: (a) test setup and (b) measurement.

Results

Half-Cell Potential Measurements

Exemplary results of the half-cell potential measurements are shown in Table 3. The results of the half cell potential measurements were plotted on a contour map for visual interpretation (Figure 6). It is evident that potential E_{corr} is low (<-350 mV) in the area around measuring points 7 to 9 from E to I, which indicates a 95% probability of corrosion. In the other measuring points, potential E_{corr} is high ($-350\,mV \leq E_{corr} \leq -200\,mV$), which indicates a 10% or uncertain probability of corrosion.

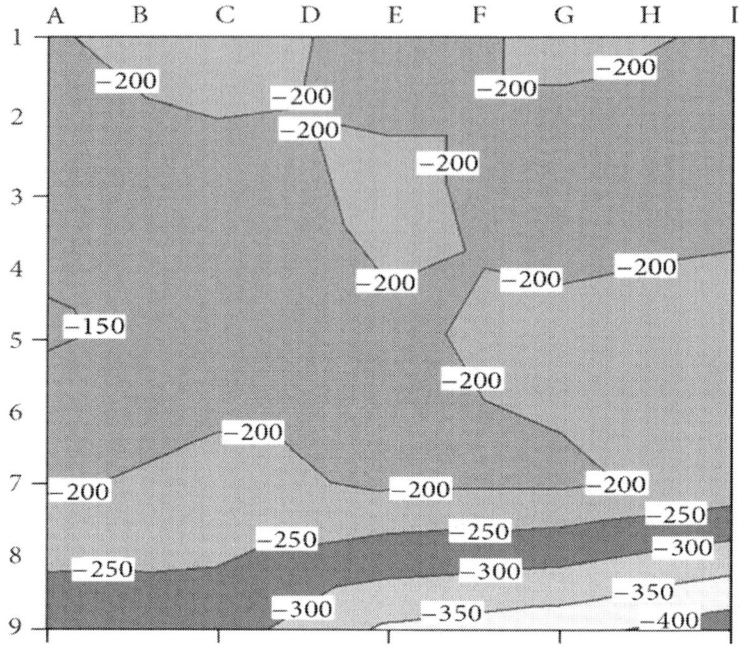

Figure 6: Contour plot of potential.

Concrete Resistivity Measurements

The results of the concrete resistivity measurements are shown in Table 4. The results of the concrete resistivity measurements were plotted on an equipotential contour map for visual interpretation (Figure 7). It is evident that concrete resistivity is low (<5 kΩcm) in the area around measuring points 5 to 9 from A to I, which indicates a very high probability of corrosion. In the other measuring points, concrete resistivity ρ is high (>5 kΩcm), indicating a high or moderate probability of corrosion.

Table 4: Concrete resistivity ρ

Nr	Concrete resistivity , kΩ cm								
	A	B	C	D	E	F	G	H	I
1	6.91	5.62	4.56	4.56	4.34	4.12	4.08	4.39	3.77

2	9.11	5.96	4.63	4.57	4.56	4.34	4.23	4.38	3.76
3	9.89	5.98	4.85	4.78	4.67	4.22	5.03	5.02	3.75
4	9.88	5.99	5.99	4.23	4.78	4.89	5.65	5.33	3.71
5	9.99	6.02	6.98	4.8	4.76	5.02	5.43	5.35	3.74
6	9.45	6.25	6.71	5.43	4.23	5.03	5.68	5.67	3.67
7	9.77	6.66	6.77	5.55	7.51	5.42	5.43	5.21	3.62
8	9.73	6.71	6.88	6.01	7.01	5.98	5.67	4.88	3.61
9	8.79	6.91	6.93	6.96	6.98	6.91	5.96	4.71	3.14

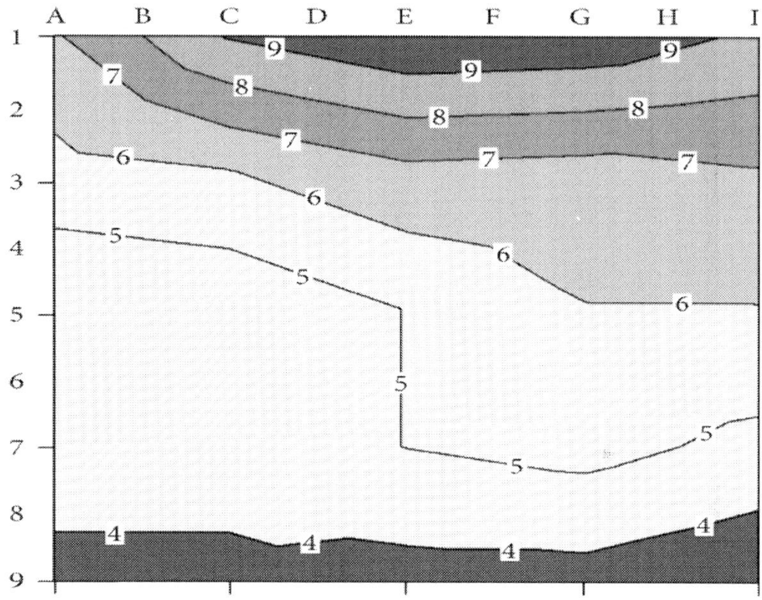

Figure 7: Contour plot of concrete resistivity ρ

Statistical Analyses of Test Results

Selected statistical characteristics of the parameters determined by half-cell potential and concrete resistivity measurements are shown in Table 5. As it appears from the histograms presented in Figure 8 and statistical characteristics shown in Table 5, the half-cell potential method yielded potential E_{corr} ranging from −432 to −144 mV. The average value of potential E_{corr} was −211.09 mV with standard deviation of 53.46 mV.

Table 5: Selected statistical characteristics of parameters determined by half-cell potential and concrete resistivity measurements

Statistical characteristics				
	Average	**Standard deviation**	**Minimum**	**Maximum**
Potential E_{corr}, mV	-211.09	53.46	-432	-144
Concrete resistivity ρ, kΩ cm	5.68	1.66	3.14	9.99

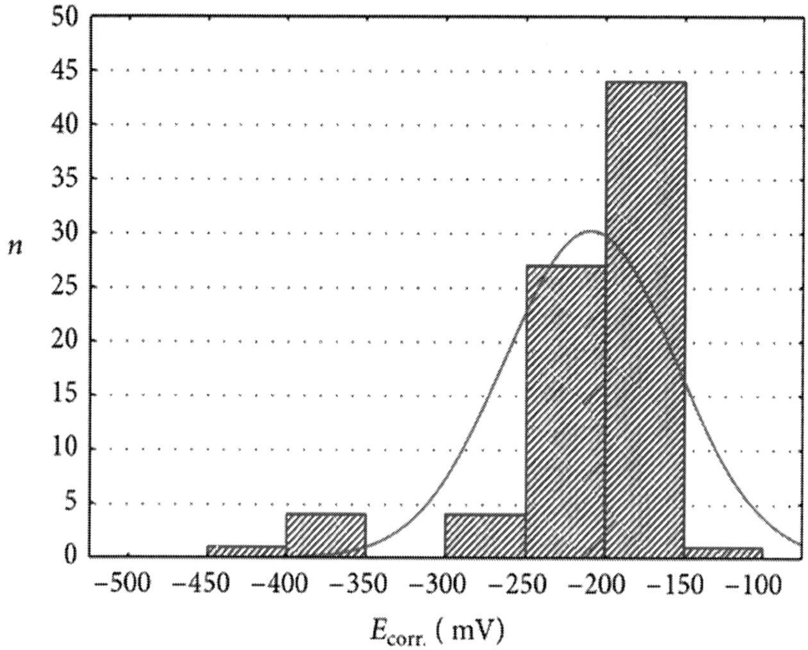

(a)

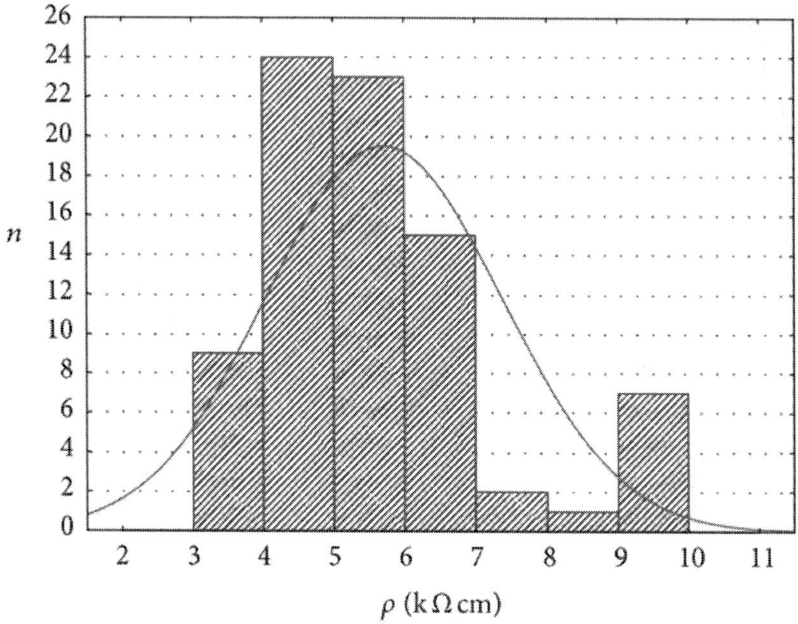

(b)

Figure 8: Histogram of (a) potential E_{corr} and (b) concrete resistivity ρ.

Concrete resistivity ρ determined by concrete resistivity measurements ranged from 3.14 to 9.99 kΩcm. The average value of concrete resistivity ρ was 5.68 kΩcm with standard deviation of 1.66 kΩcm.

Discussion

The dependence between concrete resistivity ρ and potential E_{corr}, measured on the concrete slab surface, is shown in Figure 9. One should note that potential E_{corr} sharply increases for resistivity ρ below 4 kΩcm while above 4 kΩcm, it oscillates between −150 and −250 mV.

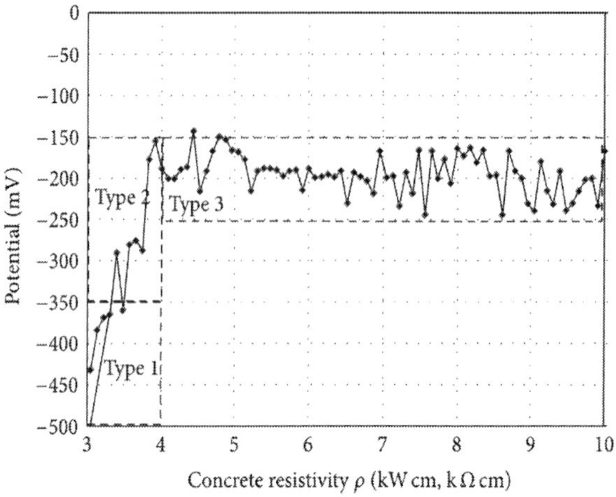

Figure 9: Concrete resistivity versus potential E_{corr}.

On the basis of Tables 1 and 2 and the results of the concrete resistivity and potential E_{corr} measurements, three areas of different types were generated (Table 6) as it has been presented in Section 2. A contourplot of the three areas is shown in Figure 10. The area around measuring points 6 to 9 from C to I is of type 1, which means that there is a 90% probability of corrosion. In the measuring points, potential E_{corr} is below −250 mV and concrete resistivity is above 4 kΩcm. The area around measuring points 3 to 9 from A to E is of type 2, which means that there is a uncertain probability of corrosion. In the measuring points, potential E_{corr} is between −150 mV and −350 mV and concrete resistivity is above 4 kΩcm. The area around measuring points 1 to 7 from A to I is of type 3, which means that there is an 10% probability of corrosion. In the measuring points, potential E_{corr} is above −250 mV and concrete resistivity is above 5 kΩcm.

Table 6: Proposed types of corrosion probability

Potential E_{corr}, mV			
	E_{corr} < −350	$-350 \leq E_{corr} \leq$ −200	$-200 \leq E_{corr}$

Concrete resistivity ρ, kΩ cm	$\rho < 4$	Type 1	Type 2	Type 2
	$4 < \rho < 5$			Type 3
	$\rho > 5$			Type 3

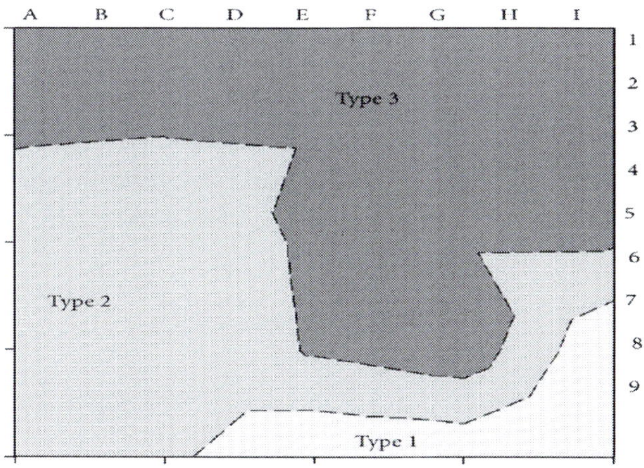

Figure 10: A contour plot of three areas of different types.

CONCLUSIONS

A methodology for assessing the probability of corrosion in concrete slabs based on a combination of two nondestructive methods, that is, the half-cell potential method and the concrete resistivity method, was briefly described. Comparative tests were carried out using the two methods to determine the probability of corrosion in model test concrete slab specimens. The experimental results showed that the two nondestructive techniques can be used together in order to obtain maximum information about the probability of corrosion in a tested structure.

This study was motivated by the engineer's need for a combination of the half-cell potential mapping technique and concrete resistivity measurements to more accurately assess the probability of corrosion. The combined techniques can be used in both the field and the laboratory environment. Moreover, they can be automated and integrated into monitoring systems for new or existing reinforced concrete structures. However, it is still recommended to perform additional tests for other rebar diameters, different aggregate grading, and a wider range of covers.

ACKNOWLEDGMENTS

This research was carried out in collaboration with the Construction and Infrastructure Group in the University of Liverpool under the project "Modern Nondestructive Testing Methods to Determine the Corrosion Rate in Concrete Structures" funded by the Erasmus LLP Programme.

REFERENCES

1. J. Berger, S. Bruschetini-Ambro, and J. Kollegger, "An innovative design concept for improving the durability of concrete bridges," Structural Concrete, vol. 12, no. 3, pp. 155–163, 2011.
2. M. Beck, A. Burkert, J. Harnisch et al., "Deterioration model and input parameters for reinforcement corrosion," Structural Concrete, vol. 13, no. 3, pp. 145–155, 2012.
3. A. Lowinska-Kluge and T. Blaszczynski, "The influence of internal corrosion on the durability of concrete," Archives of Civil and Mechanical Engineering, vol. 12, no. 2, pp. 219–227, 2012.
4. K. Osterminski and P. Schießl, "Design model for reinforcement corrosion," Structural Concrete, vol. 13, no. 3, pp. 156–165, 2012.
5. H. Song and V. Saraswalhy, "Corrosion monitoring of reinforced concrete structures—a review,"International Journal of Electrochemical Science, vol. 2, pp. 1–28, 2007.
6. C. Andrade, C. Alonso, J. Gulikers et al., "Test methods for on-site corrosion rate measurement of steel reinforcement in concrete by means of the polarization resistance method," Materials and Structures, vol. 37, no. 273, pp. 623–643, 2004.
7. G. Sposito, P. Cawley, and P. B. Nagy, "Potential drop mapping for the monitoring of corrosion or erosion," NDT and E International, vol. 43, no. 5, pp. 394–402, 2010.

8. "ASTM C876-91: standard test method for half-cell potentials of uncoated reinforcing steel in concrete," 1999.
9. J. Hoła and M. Ksiazek, "Research on usability of sulphur polymer composite for corrosion protection of reinforcing steel in concrete," Archives of Civil and Mechanical Engineering, vol. 9, no. 1, pp. 47–59, 2009.
10. M. Kosior-Kazberuk and W. Jezierski, "Evaluation of concrete resistance to chloride ions penetration by means of electric resistivity monitoring," Journal of Civil Engineering and Management, vol. 11, no. 2, pp. 109–114, 2005.
11. J. Hoła and K. Schabowicz, "State-of-the-art non-destructive methods for diagnostic testing of building structures—anticipated development trends," Archives of Civil and Mechanical Engineering, vol. 10, no. 3, pp. 5–18, 2010.
12. K. Liam, S. Roy, and D. Northwood, "Chloride ingress measurements and corrosion potential mapping study of a 24-year-old reinforced concrete jetty structure in a tropical marine environment research,"Magazine of Concrete Research, vol. 44, pp. 205–215, 1992.
13. J. Flis, H. W. Pickering, and K. Osseo-Asare, "Assessment of data from three electrochemical instruments for evaluation of reinforcement corrosion rates in concrete bridge components," Corrosion, vol. 51, no. 8, pp. 602–609, 1995.
14. M. G. Grantham, B. Herts, and J. Broomfield, "The use of linear polarisation corrosion rate measurements in aiding rehabilitation options for the deck slabs of a reinforced concrete underground car park," Construction and Building Materials, vol. 11, no. 4, pp. 215–224, 1997.
15. V. Živica, "Possibility of improvement of potentiodynamic method for monitoring corrosion rate of steel reinforcement in concrete," Bulletin of Materials Science, vol. 24, no. 5, pp. 555–558, 2001.
16. S. Erdogdu, I. L. Kondratova, and T. W. Bremner, "Determination of chloride diffusion coefficient of concrete using open-circuit potential measurements," Cement and Concrete Research, vol. 34, no. 4, pp. 603–609, 2004.
17. N. Bowler and Y. Huang, "Electrical conductivity measurement of metal plates using broadband eddy-current and four-point

methods," Measurement Science and Technology, vol. 16, no. 11, pp. 2193–2200, 2005.

18. T. Gorzela czyk, "Moisture influence on the failure of self-compacting concrete under compression,"Archives of Civil and Mechanical Engineering, vol. 11, no. 1, pp. 45–60, 2011.
19. R. Polder, C. Andrade, B. Elsener et al., "Test methods for on site measurement of resistivity of concrete," Materials and Structures, vol. 33, pp. 603–611, 2000.
20. B. Elsener, L. Bertolini, P. Pedeferri, and R. Polder, Corrosion of Steel in Concrete—Prevention, Diagnosis, Repair, Wiley-Vch, Berlin, Germany, 2004.
21. S. Feliu, J. A. Gonzalez, S. Feliu, and M. C. Andrade, "Confinement of the electrical signal for in situ measurement of polarization resistance in reinforced concrete," ACI Materials Journal, vol. 87, no. 5, pp. 457–460, 1990.
22. G. K. Glass, C. L. Page, and N. R. Short, "Factors affecting the corrosion rate of steel in carbonated mortars," Corrosion Science, vol. 32, no. 12, pp. 1283–1294, 1991.
23. W. López, J. A. González, and C. Andrade, "Influence of temperature on the service life of rebars,"Cement and Concrete Research, vol. 23, no. 5, pp. 1130–1140, 1993.
24. W. Morris, A. Vico, M. Vazquez, and S. R. de Sanchez, "Corrosion of reinforcing steel evaluated by means of concrete resistivity measurements," Corrosion Science, vol. 44, no. 1, pp. 81–99, 2002.
25. J. Zhang, P. J. M. Monteiro, and H. F. Morrison, "Noninvasive surface measurement of corrosion impedance of reinforcing bar in concrete—part 1: experimental results," ACI Materials Journal, vol. 98, no. 2, pp. 116–125, 2001.
26. J. Zhang, P. J. M. Monteiro, H. F. Morrison, and M. Mancio, "Noninvasive surface measurement of corrosion impedance of reinforcing bar in concrete—part 3: effect of geometry and material properties,"ACI Materials Journal, vol. 101, no. 4, pp. 273–280, 2004.
27. B. Elsener, C. Andrade, J. Gulikers, R. Polder, and M. Raupach, "Half-cell potential measurements—potential mapping on

reinforced concrete structures," Materials and Structures, vol. 36, no. 261, pp. 461–471, 2003.

28. L. Sadowski, "New non-destructive method for linear polarisation resistance corrosion rate measurement," Archives of Civil and Mechanical Engineering, vol. 10, no. 2, pp. 109–116, 2010.
29. S. Millard and L. Sadowski, "Novel method for linear polarisation resistance corrosion measurement," e-Journal of Nondestructive Testing & Ultrasonics, vol. 14, 2009.
30. R. Vedalakshmi, L. Balamurugan, V. Saraswathy, S. H. Kim, and K. Y. Ann, "Reliability of Galvanostatic Pulse Technique in assessing the corrosion rate of rebar in concrete structures: laboratory vs field studies," KSCE Journal of Civil Engineering, vol. 14, no. 6, pp. 867–877, 2010.
31. J. Rhazi, "Test method for evaluating asphalt-covered concrete bridge decks using ground penetrating radar," in PIERS Proceedings, Marrakesh, Morocco, 2011.
32. J. Bungey and S. Millard :, Testing of Concrete in Structures, Chapman & Hall, Glasgow, UK, 1996.
33. E. Bardal and J. Drugli :, "Corrosion detection and diagnosis," in Materials Science and Engineering, Vol. 3, Encyclopedia of Life Support Systems, R. D. Rawlings, Ed., 2004.
34. W. Elkey and E. Sellevold, Electrical Resistivity of Concrete, Publication no. 80, Norwegian Road Research Laboratory, Oslo, Norway, 1995.
35. L. Sadowski, "Non-destructive investigation of corrosion current density in steel reinforced concrete by artificial neural networks," Archives of Civil and Mechanical Engineering, vol. 13, no. 1, pp 104–111, 2013

Chapter 5

Impact of Underground Mining on Shaft Lining and Aquifer in Eastern China

Qing Yu, Hideki Shimada, Takashi Sasaoka, and Kikuo Matsui

Department of Earth Resources Engineering, Faculty of Engineering, Kyushu University, Fukuoka, Japan

ABSTRACT

Serious shaft lining ruptures have often occurred in the eastern part of China since 1987 due to the complicated geological conditions. This paper tries to find out the relationship between mechanisms of shaft lining rupture and the underground mining process. The analysis is based on the existence typical engineering and geological conditions in eastern China; the impact of underground mining on the shaft lining and aquifer layer is analyzed by using numerical method. The impact factors such as different depths, thicknesses, mining widths of coal seam and different distances to the shaft are used in the analysis. The mining area under the aquifer which near the shaft lining has a significant impact on the shaft lining due to mining process, and increases the risk of occurrence of shaft lining rupture.

INTRODUCTION

Serious shaft lining ruptures have often occurred in the eastern part of China and this problem has been one of the major subjects in Chinese coal mine industry for past twenty years. All these sites that shaft lining rupture occurred are in similar geological conditions. All these shafts pass through deep alluvium of Quaternary strata which the composition of the bottom aquifer is complex [1].

According to the statistical data of the shaft lining rupture in alluvium, the elapsed time after drivaging and location of rupture has its own rules. Generally, the type of shaft lining ruptures is divided into three phases [2]: 1) Rupture during the freezing shaft sinking: horizontal cracks caused by temperature variation is initiated and propagated; 2) Frozen wall melting: horizontal and inclined cracks caused by temperature variation are initiated and propagated; 3) Surface subsidence after several years: horizontal and inclined cracks caused by surface subsidence are propagated, especially, this phenomena is occurred near the boundary between aquifer and bedrock.

The shaft lining rupture happening after several years running is the most common situation in the eastern part of China, such as Datun, Xuzhou, Huaibei, Yanzhou, Yingxia, Hebi, Dongrong, etc., since 1987 [3-5]. When shaft lining rupture occurs, the inner shaft lining delaminates and spalls, longitudinal steel bows inward, transverse cracks form and intersect in the horizontal direction along circle, seepage occurs or even sand gushes, and mostly seriously, concrete blocks fall out and break the equipment in shaft. In addition, the shaft bends up; cage guides, drainpipes and pressure ventilation pipes are in longitudinal bending, and in serious cases the cage is stuck due to torsional deformation.

According to several surveys on coal mines, before the shaft lining rupture happening, the groundwater leakage has been observed in the mining area from roof, the groundwater level decline is happening at the same time in the aquifer [6-9]. The current researches are mostly focused on the causes, mechanisms and solutions for these kinds of geotechnical issues and they have been the major topics in recent 20 years. In this research, the model of strata layers is built and used for numerical simulation, and the impact of coal mining in the deeper coal seam before aquifer drawdown on shaft lining is examined.

ANALYSIS MODEL

FLAC^{3D} is a three-dimensional explicit finite-difference program for engineering mechanics computation and simulating the behavior of three-dimensional structures built of soil, rock or other materials that undergo plastic flow when their yield limits are reached.

The basic model used in the analysis is shown in Figures 1 and 2. This model represents the typical engineering and geological conditions in eastern China especially Baodian coal mine which happened the shaft lining rupture in 1995. The net diameter of the main shaft in this mine is 6.5 m and the thickness of the shaft lining is 1.0 m constructed by C28 concrete. The thickness of alluvium is 148.69 m and it contains 3 aquifers. The bottom aquifer is important to the mine and the water level is affected by mining, so in this model, only the bottom aquifer is left for simulation.

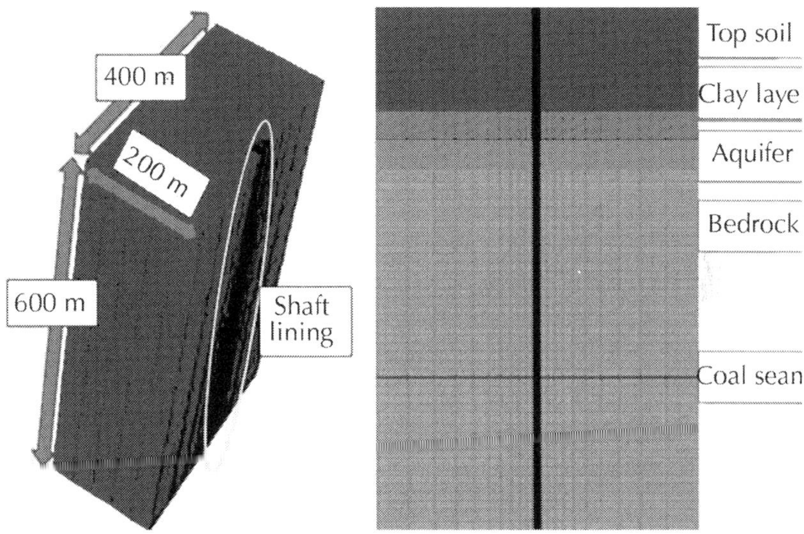

Figure 1: Analysis model.

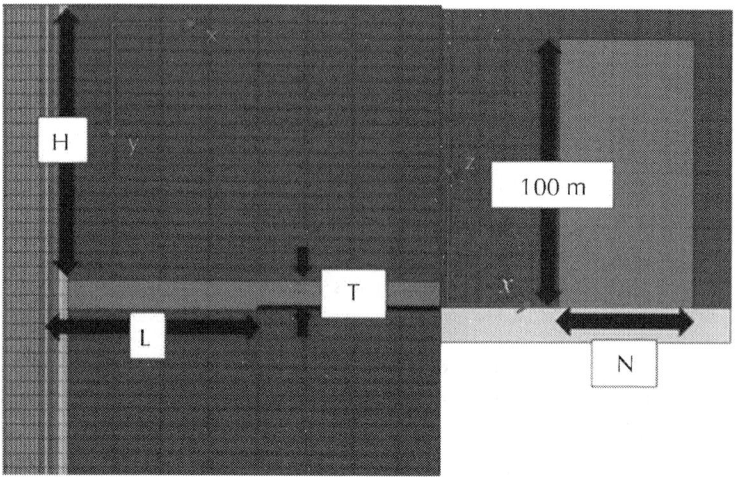

Figure 2: Mining area model.

According to the engineering geological data and shaft construction data of coal mine, the required mechanical parameters for numerical simulation are shown in Table 1.

As the process of soil deformation and shaft lining stress change is very complex, the following assumptions are made:

- Shaft lining, soil and loss areas are treated as the symmetrical distribution, belong to the space axial symmetry.
- Concrete shaft lining, the surrounding soil and aquifer are homogeneous, isotropic. No interface between the surrounding soil and shaft lining, that means no sliding in the interface.

Then the boundary conditions are set as follows: a free top surface, two horizontal fixed slides and a vertical fixed the lower surface.

As the coal seam is under the deep bottom aquifer, the different depth, thickness, mining width of coal seam and different distance to the shaft were simulated as the impact factors of the shaft lining and the aquifer (as shown in Table 2).

ANALYSIS RESULTS AND DISCUSSION

Failure Initiation Due to Mining Process

The shear strain change is chosen to reveal the impact of the mining process, as the rock compression failure mainly belongs to the shear failure [10]. Figure 3 shows the shear strain distribution around the mining area and the aquifer during mining. The high strain area is in yellow colour and the low strain area is in blue. From this figure, the both ends of the mining area and the aquifer layer have the high strain. In addition, the shear strain in the rock layer between the mining area and the aquifer exceeds 2.0×10^{-3}. And except the part shown in blue, there is a risk of failure occurs, especially the area between mining area and aquifer. Therefore, the leakage occurs at the impermeability rock layer can be concluded by the fissure developing in the rock which the failure happened during the mining process. According to the above analysis, during the underground mining in 250 m depth, the bedrock under aquifer begins to get loose and then some fissure develops, as a result the aquifer drainage occurs.

Figures 4 and 5 show the distribution of shear strain in different mining depth. The deeper of the mining area, the distance to the aquifer is greater, then the impact on the aquifer is gradually reduced. The impact on the aquifer becomes disappear while mining in 400 m depth, and the shear strain near the mining area tends to become higher. For this reason, if the mining depth is deeper than 400 m, it becomes much more important to consider the safety of the surrounding area compare with the impact on the aquifer.

Table 1: Mechanical parameters of soil and shaft lining

Material	Top soil	Clay	Aquifer	Bedrock	Shaft lining	Coal seam
Thickness (m)	120	30	30	450	1	
Dry density (g/cm^3)	1.65	1.65	1.7	2.7	3.0	2.7

Bulk modulus (Mpa)	50	50	61.25	1.1 x 10^4	2 x 10^4	1.1 x 10^4
Shear modulus (Mpa)	23	23	28	8333	12600	8333
Porosity	0.18	0.05	0.3			

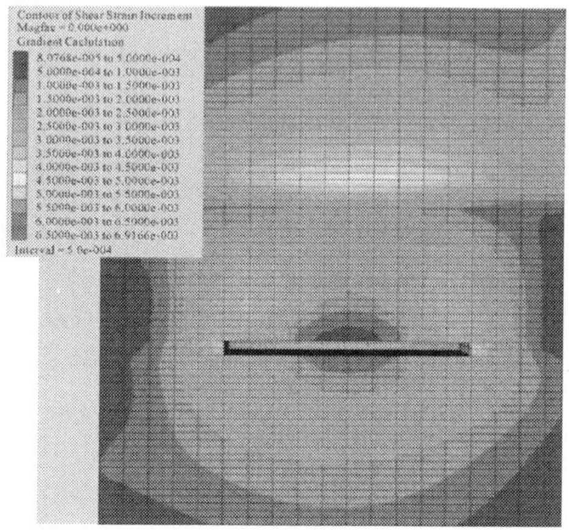

Figure 3: Shear strain distribution during mining (50 m to the shaft and 250 m mining depth, 6 m thickness coal seam).

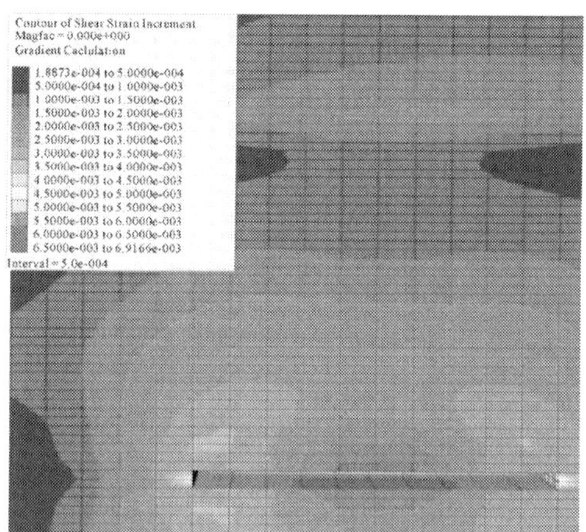

Figure 4: Shear strain distribution during mining (50 m to the shaft and 300 m mining depth, 6 m thickness coal seam).

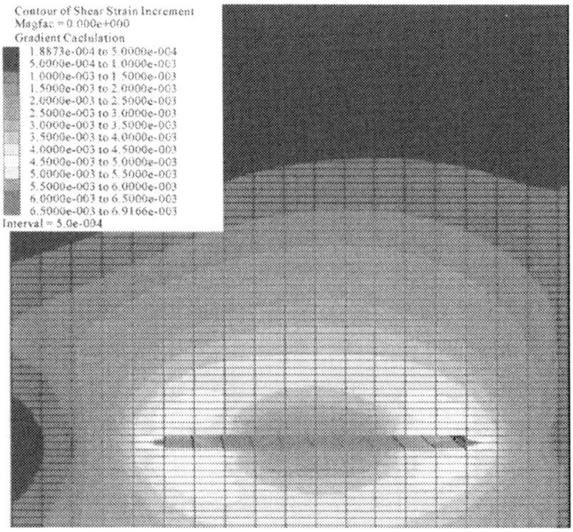

Figure 5: Shear strain distribution during mining (50 m to the shaft and 400 m mining depth, 6 m thickness coal seam).

Table 2: Factors for analysis

Depth of coal seam (H)	250 m x 300 m x 400 m
Thickness of coal seam (1)	3 mx 6mx 9m
Mining width (N)	50 m x 100m
Distance to shaft (L)	50 m x 70 m x 90 m

Impact Due to Mining Distance to Shaft

The stress in shaft lining is the most concerned issue, so the impact to the shaft lining is revealed by stress change. Figure 6 shows the impact on the surrounding areas and the shaft lining by different distance of mining area to the shaft. The lines in the figure are the values of maximum principal stress in the surrounding area, aquifer and shaft lining. As the distance between mining area and shaft becomes greater, the maximum principal stress in shaft lining becomes smaller, the mining impact on the shaft lining becomes smaller and that can reduce the concentration of the maximum stress.

Impact Due to Mining Depth

The mining depth also has the impact to the stress in the shaft lining, though Figure 7 shows the maximum principal stress change due to different mining depths. As the mining area is deeper and the overburden increases, the deformation of surrounding area is greater and the stress concentration in mining area and shaft lining tends to rise as the depth increases. Also the maximum principal stress in the mining area is higher than in the shaft lining. From the above analysis, the risk of shaft lining rupture is smaller than the risk of mining area collapse. So the coal pillar and artificial support need to be remained in the mining area considering the collapse risks.

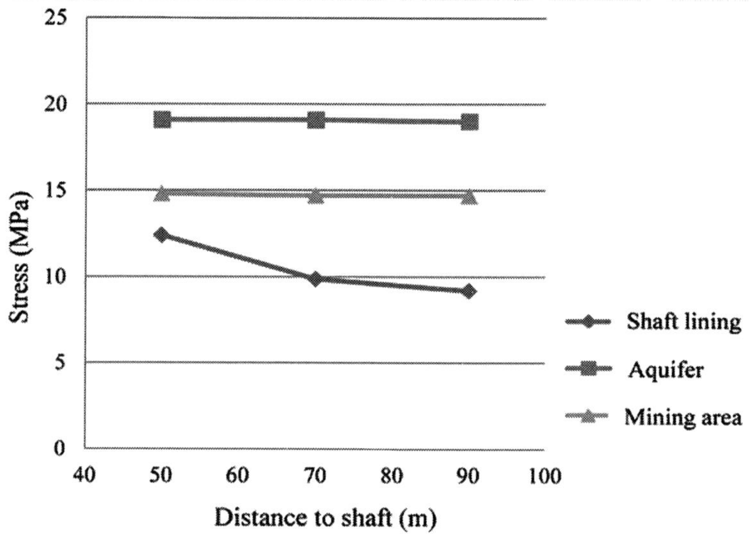

Figure 6: Impact due to distance to shaft (250 m mining depth).

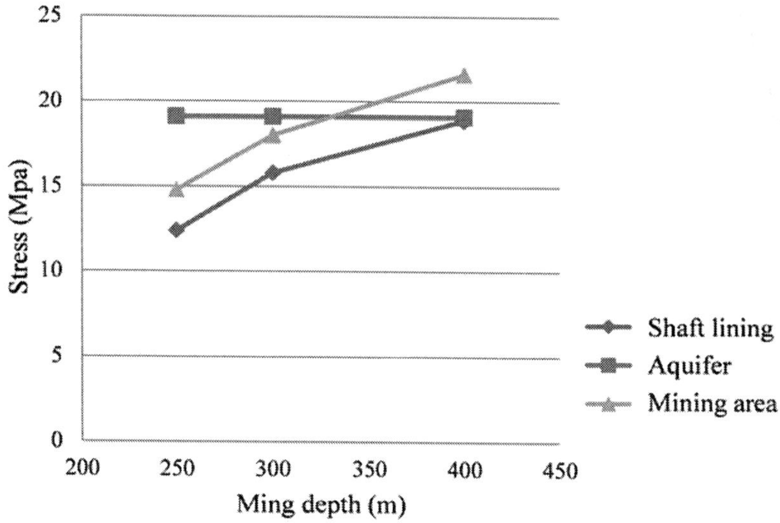

Figure 7: Impact due to mining depth (50 m to shaft).

Impact Due to Mining Width

The change of mining width leads to the deformation of the ground layers; the displacement of the ground layers is used to show the impact. Figures 8 and 9 show the influence of 50 m and 100 m mining width As the mining width increases, the deformation of surrounding mining area increases which it also gives an impact on the deformation of the shaft lining. As shown in Figure 10, when the mining width is 50 m, the maximum principal stress concentration occurs in the surrounding area, and when the mining width is 100 m, the maximum stress occurs in the shaft lining.

Therefore, due to the deformation of the mining area, there is a significant stress concentration occurs in shaft lining. Particularly, a great stress concentration has occurred around the mining area and the impact range of 50 m mining width and 100 m mining width has a large difference. From above analysis, the mining width has a significant impact on the shaft lining, and the failure in shaft lining is much more severe than in the mining area.

Impact Due to Coal Seam Thickness

Three thicknesses of coal seam are set as 3 m, 6 m and 9 m for examining the impact. And the stress change in the shaft lining, aquifer and mining area is used to reveal the impact of the coal seam thickness.

Figure 11 shows the maximum principal stress in the mining area, aquifer and shaft lining. From the figure, the stress in aquifer has no obvious change. However, the stress in mining area has a significant increase with the increase of the coal seam thickness. Furthermore, the stress in mining area has outstripped the stress in shaft lining. That means the thickness change of coal seam has an impact on the mining area, but not obvious in the shaft lining and aquifer.

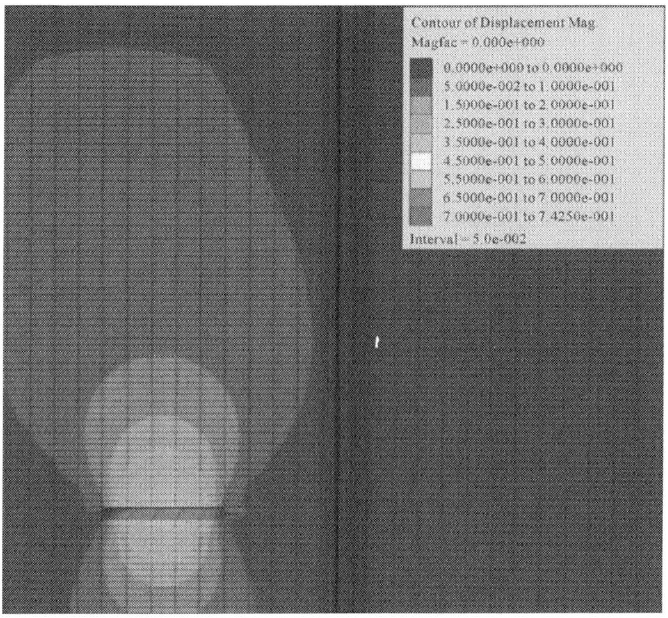

Figure 8: Vertical displacement distributions under 50 m mining width (50 m to the shaft and 400 m mining depth, 6 m thickness coal seam).

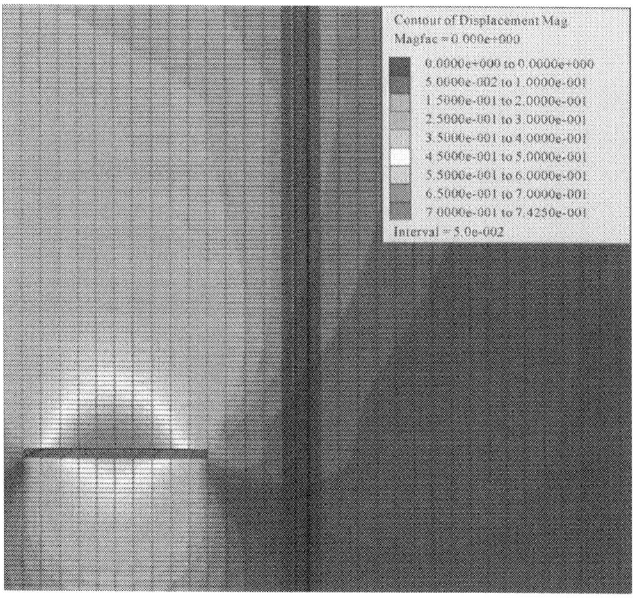

Figure 9: Vertical displacement distributions under 100 m mining width (50 m to the shaft and 400 m mining depth, 6 m thickness coal seam).

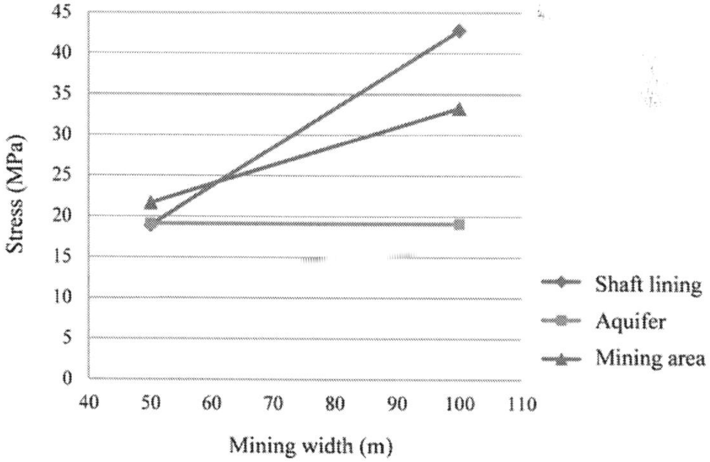

Figure 10: Impact due to mining depth due to mining width (50 m to the shaft and 400 m mining depth, 6 m thickness coal seam).

CONCLUSIONS

In this study, the impact of the underground mining on the shaft lining and aquifer is analyzed by numerical simulation. The main aim is to find out the relationship between mechanisms of shaft lining rupture and the underground mining process. The impact factors such as different depth, thickness, mining width of coal seam and different distance to the shaft are used in the analysis. From a series of numerical simulations, the impact of different factors can be cleared as follows:

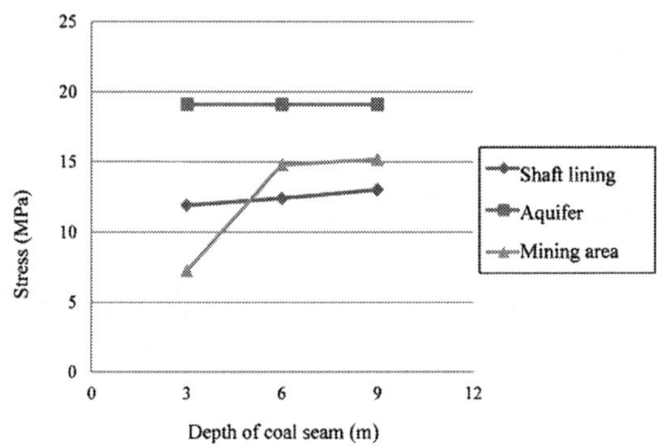

Figure 11: Impact due to coal seam thickness (50 m to shaft and 250 m mining depth).

- The underground mining process can lead to the initiation of the failure in the soil and rock layers. If the mining location is near some certain range of the aquifer, the aquifer drainage will occur due to the fissure in aquifer connection to fissures in the bedrock. This situation should be avoided in the mining process by changing the mining location or using supporting structure.
- The impact of the mining distance to the shaft lining is obvious, if the mining location is closed to the shaft lining, the greater stress concentration will happen in shaft lining and increase the risk of occurrence of shaft lining rupture.
- When the mining location is near the shaft ling in a certain range, the increase of the mining depth also can increase the risk of occurrence of shaft lining rupture. More seriously, the possibility of mining area collapse is raised due to the increase of mining

depth.

- The mining width has a significant impact on the shaft lining, the stress in shaft lining increase fast due to the increase of mining width. Before the increase of the mining width, the support structure should be prepared to protect the mining area in case the stress transfers to shaft lining.
- The thickness change of coal seam has an impact on the mining area, but not obvious in the shaft lining and aquifer.
- Therefore, when the mining area under the aquifer nears the shaft lining, there is a significant impact on the shaft lining due to mining process, and the risk of occurrence of shaft lining rupture become higher and need to be concerned.

ACKNOWLEDGEMENTS

Authors are grateful for financial assistance provided by the Global-Centre of Excellence in Novel Carbon Resource Science, Kyushu University.

REFERENCES

1. G. Cui and X. Cheng, "Occasions of Damaging Shaft Walls in Xuhuai District," Coal Science and Technology, Vol. 19, No. 8, 1991, pp. 46-50.
2. L. Jing and H. Wang, "The Research about the Fracture Regularity of the Frozen Shaft Wall in Surface Soil," Journal of Huainan Institute of Technology, Vol. 20, No. 1, 2000, pp. 14-20.
3. M. Gu, "Failure of Shaft Caused by Special Geologic Hazards and Corresponding Prevention Measures," Journal of Engineering Geology, Vol. 8, No. 2, 2000, pp. 197-201.
4. W. Yang and H. Fu, "Theoretical Investigation on Vertical Additional Force on Shaft Lining in Special Stratum," Journal of China University of Mining & Technology, Vol. 9, No. 2, 1999, pp. 129-135.
5. G. Zhou and G. Cui, "Numerical Analysis on Interaction between Shaft Wall and Surrounding Strata after Aquifer Grouting," Journal

of China University of Mining & Technology, Vol. 27, No. 2, 1998, pp. 135-139.

6. H. Liu, W. Chen and Z. Wang, "Theoretical Analysis of Shaft Lining Damage Mechanism of Yanzhou Mine," Chinese Journal of Rock Mechanics and Engineering, Vol. 26, Supp. 1, 2007, pp. 2620-2626.
7. S. Bi, X. Lou and B. Xu, "On the Mechanism of Coal Mine Shaft Damage Caused by Subsidence in Xuhuai Area, Southeast China," Communications in Nonlinear Science & Numerical Simulation, Vol. 2, No. 2, 1997, pp. 75-80. doi:10.1016/S1007-5704(97)90043-5
8. Yang W., Cui G. and Zhou G., "Fracture Mechanism of Shaft Lining under Special Strata Condition and the Technique Preventing the Shaft From Fracturing (Part One)," Journal of China University of Mining & Technology, Vol. 25, No. 4, 1997, pp. 1-4.
9. G. Zhou and X. Cheng, "Study on the Stress Calculation of Shaft Lining Surrounded by Special Strata," Journal of China University of Mining &Technology, Vol. 24, No. 4, 1995, pp. 24-30.
10. M. You and A. Hua, "Fracture of Rock Specimen and Decrement of Bearing Capacity in Uniaxial Compression," Chinese Journal of Rock Mechanics and Engineering, Vol. 17, No. 3, 1998, pp. 292-296.

Chapter 6

Survey on Storage of Water for Domestic Use in Underground Reinforced Concrete Tanks: The Case of Brazzaville (Republic of Congo)

Narcisse Malanda and, Paul Louzolo-Kimbembe

Laboratory of Civil Engineering and Environmental Engineering, National School of Engineering, University Marien Ngouabi, Brazzaville, Congo

ABSTRACT

The problem of water supply from the public distribution network still poses very serious problems in many cities in developing countries. Intermittent water supply pushes some households to build underground reinforced concrete tanks for water storage to cope with unwanted water breaks. This study that relies on the results of a survey of households in some areas of the city of Brazzaville (Republic of Congo) aims to verify the importance that users attach to the quality of the works constructed. Indeed, the reliability of the tanks resulting in their impermeability to the external environment has a direct impact on the quality of stored water and therefore the use that is made of water daily. Five areas were selected because of their soil moisture. By 256 tanks identified, 143 are made of reinforced concrete and 113 in masonry. The coating materials used to seal the walls are preferably the earthenware tiles (64% of tanks), then the Sika cement (31%). Food painting (5%) is only rarely used. However, 66% of households are not assured of the portability of the water stored. A significant number of households (46%) think that the stored water could be contaminated with noxious substances seeping from the outside through the walls of the tanks. The issue of sealing of underground water tanks, especially in areas where the water table is shallow, seems concerned users.

INTRODUCTION

In many cities in developing countries, the permanent water supply from the public distribution network is not always adequately insured [1]. It under goes regular water breaks. In the city of Brazzaville (Republic of Congo), to overcome the inconvenience caused by the intermittent supply, many households and especially commercial establishments have opted for a system of water storage in underground tanks, built in reinforced concrete or in masonry. The water is then sent into the building using a suppressor. Also, the storage of water for domestic use in underground tanks, especially in areas with moisture saturated soil and polluted soil, is an interesting issue. It appears that little work has been done in this context. Most studies encountered mainly interested in the analysis of surface water and groundwater [2], hydro chemical assessment, characterization of the quality and Hydro geochemistry of groundwater [3]-[6].These studies examine the review of hydro chemical parameters with the help of statistical tools. In our study, it is not about groundwater, but about water stored in underground tanks built in reinforced concrete or in masonry, with rectangular form. The water is collected directly from the network of the National Water Distribution Company (SNDE) or sometimes by buying from private suppliers who deliver with tankers. If the quality of the water collected is not questioned, however, the quality of water stored during a certain time in the tank should be examined. While sealing the walls of the water tank is not effective, there is a real risk of contamination of water by transfer of pathogens from the external environment. In this study, we try to know, at first, the distribution of water tanks in some areas of the city of Brazzaville. Then, using a questionnaire, we collect information from households to assess the value that they place on the quality of water tanks, because the portability of the stored water is closely linked to their construction processes.

METHODOLOGY

A survey was conducted in Brazzaville in October and November 2012, to collect data from users of underground water tanks.

Having established the questionnaire, the method used for data collection was the census, that is to say the passage of investigators in all the streets and all the parcels areas previously targeted as study areas.

This study focused on three main modules:

- Identification of water tanks;
- Knowledge of materials of realization of the tanks;
- Knowledge of portability of water.

The results of the survey are presented in a summary table in which you can make a holistic approach to all areas and a refined approach to each area with the main indicators presented.

The table of survey results is followed by an analytical section.

Note that the survey was conducted thanks to the perspicacity of investigators who, by direct observation, were able to compensate for the reluctance of the water tank users most often suspicious rather than cooperative.

RESULTS AND DISCUSSION

Results and Analysis

We present first the limitation of survey areas and the number of tanks per area in Table 1.

Data analysis on the results of the survey will be done by considering modules and answers to questions, expressed in absolute numbers and percentages. The following three modules were chosen:

- Identification;
- Knowledge of building materials;
- Knowledge of water portability.

Identification

For the identification module, we chose the cutting of areas relative to the relief and habits of people to use as water tanks and water wells. We found low areas watered by small streams, such as Madoukou and La Tsiémé (Figure 1).

In these areas, the water table is shallow, which tempts people to want to dig wells. On the other hand, given the difficulties in the regular supply of water by the National Water Distribution Company (SNDE), some residents of Brazzaville decided to build underground water tanks.

Table 1: Overview of survey areas and identification of water underground tanks

Survey areas	Limitation of survey areas	Observations
Area A	Tsiémé Market until Jeanne Vialle, Madoukou river, avenue de France.	3 tanks in masonry and 21 reinforced concrete tanks identified

Area B	Loutassi avenue; Madoukou river—10 Francs market; Madoukou and La Mfoa rivers; City of Poto-Poto.	37 tanks in masonry and 61 reinforced concrete tanks identified, with 9 wells.
Area C	Itoumbi-Ouenzé; Avenue de La Tsiemé—La Tsiemé market; Boueta Mbongo street—La Tsiemé river.	12 tanks in masonry and 40 reinforced concrete tanks, with 52 wells were identified.
Area D	Itoumbi roundabout (Cité des 17 avenue); Itoumbi street; BouetaMbongo avenue; Moukondo Market.	30 tanks in masonry and 15 reinforced concrete tanks plus 8 wells identified.
Area E	La Tsiémé cemetery—La Tsiémé river; Mikalou—High School Thomas Sankara ; Talangaï hospital.	31 tanks in masonry and 6 reinforced concrete tanks.
Total all areas	256 water tanks including: • 113 tanks in masonry; • 143 reinforced concrete tanks	

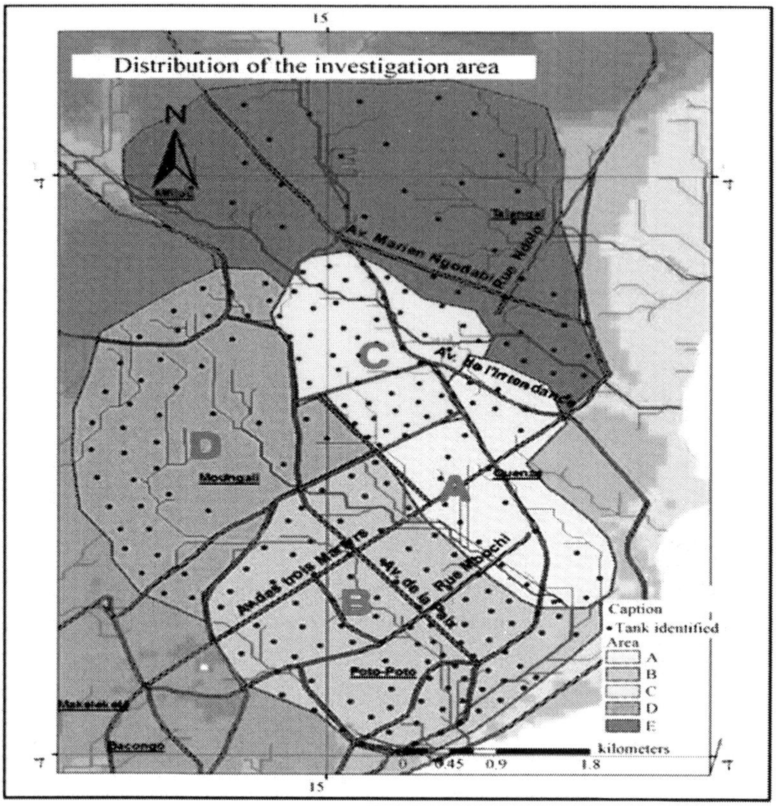

Figure 1: Cartographic representation of investigation areas. Dots (·) represent geographical position of water tanks.

Note that 256 water tanks in total were identified in localized areas (Table 1). They are unevenly distributed (Figure 2). This is the area B which has the highest number with 98 tanks (38.3%), followed by the area C with 52 tanks (20.3%). The area A has the lowest number with 24 tanks (9.4%).

Knowledge of Materials Used for the Construction of Water Tanks

To better appreciate the importance of the issue, it is necessary to know the type of materials used by households for the construction of water tanks.

Overall, it appears that households build their water tanks preferably in reinforced concrete (56%), the construction in masonry representing only 44% of the number of tanks identified in all areas (Figure 3).

We can also compare the use of materials in relation to the areas (Figure 4). If in areas A, B and C the use of reinforced concrete is paramount with 8.2%, 23.8% and 15.6% respectively of households, it should be noted that the construction in masonry prevalent in areas D and E with 11.7% and 12% respectively. However, both types of materials are used in all areas, at various degrees (Table 2).

Regarding the type of materials that were used to seal tanks identified, earthenware tiles are widely preferred with 64.1%, followed by Sika cement and food paint with 30.8% and 5.1% respectively (Table 3). In Figure 5 we show the use of sealing materials per area. We show in Table 4 the frequency of use of sealants.

Overall, the tiles are much more used to seal both the reinforced concrete tanks and the masonry in masonry. Sika cement is mainly used for reinforced concrete tanks. Food paint has been selected by a small number of users (Figure 6).

The trade that has more demanded for the realization of the tanks is the worker with 44.5% of the tanks constructed. However more than half of the tanks were realized by skilled technicians including work technicians with 34% and engineers with 21.5% (Figure 7).

We find that the most the qualification is higher; less one is called to build tanks. This is certainly due to the much higher fees charged by technicians and engineers. In all areas, engineers are the less demanded for construction of water tanks. The workers come in head in areas B, D and E, while technicians dominate in areas A and C (Figure 8).

Table 2: Results of the survey on water tanks in Brazzaville

	Questions Answers Survey area	Water tank?	Information about the carrying out of tanks?		Nature of the materials used to build the tanks?		Qualifying trades who carried out the work?		
		Yes	Yes	No	Masonry	Reinforced Concrete	Engineer	Technician	Worker
Area A	Number of tanks	24	13	11	3	21	4	12	8
	Percentage (relative to the total)	9.4%	5%	4.3%	1.1%	8.2%	1.6%	4.7%	3.1%
Area B	Number of tanks	98	51	47	37	61	24	35	39
	Percentage (relative to the total)	38.3%	20%	18.4%	14.5%	23.8%	9.4%	13.7%	15.3%
Area C	Number of tanks	52	18	34	12	40	13	20	19
	Percentage (relative to the total)	20.3%	7%	13.3%	4.7%	15.6%	5%	7.8%	7.5%

Area D	Number of tanks	45	22	23	30	15	12	11	22
	Percentage (relative to the total)	17.6%	8.6%	9%	11.74	6%	4.7%	4.3%	8.6%
Area E	Number of tanks	37	24	13	31	6	2	9	26
	Percentage (relative to the total)	14.4%	9.4%	5%	12%	2.4%	0.8%	3.5%	10%
Total all areas	Number of tanks	256	128	128	113	143	55	87	114
	Percentage	100%	50%	50%	44%	56%	21.5%	34%	44.5%

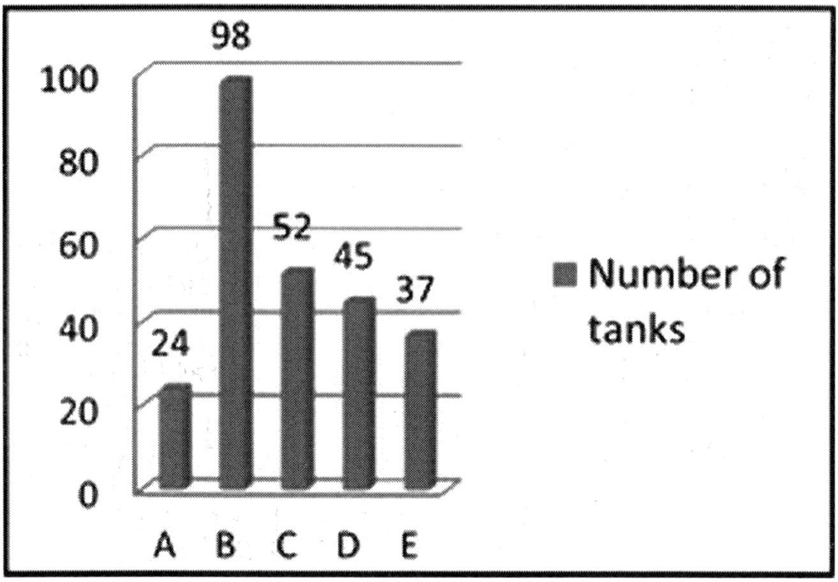

Figure 2: Distribution of tanks per area.

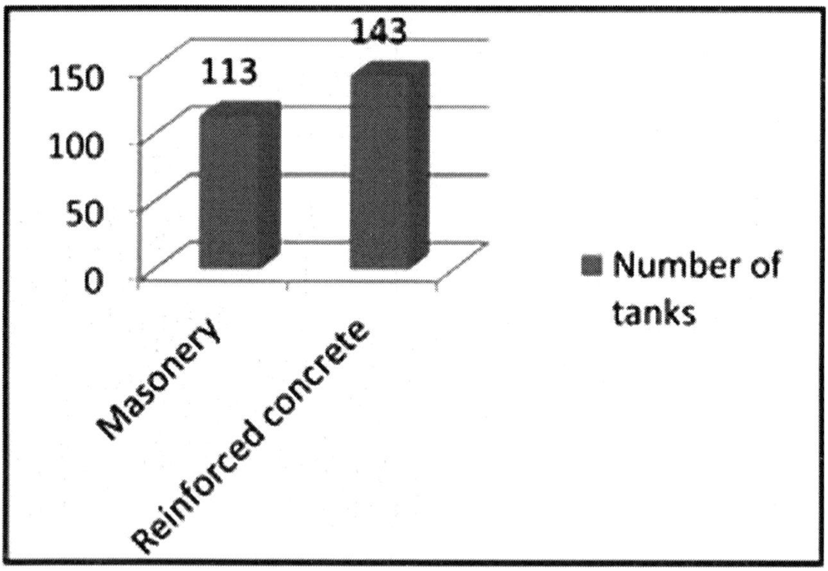

Figure 3: Type of materials used for the construction of water tank.

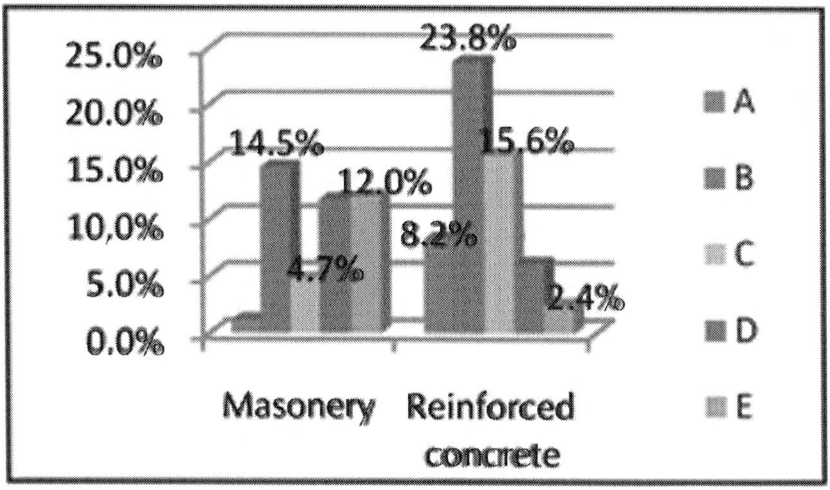

Figure 4: Sort of materials used for the construction of water tank per area.

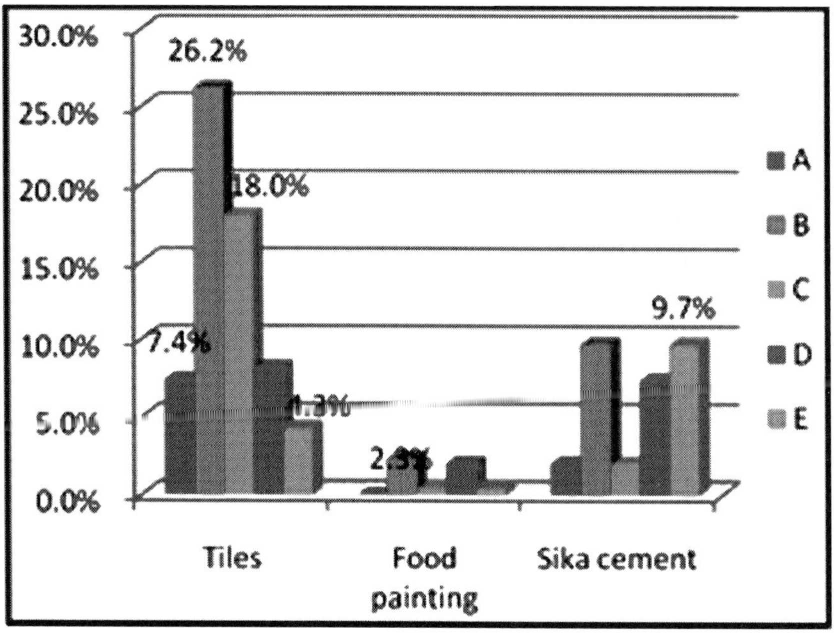

Figure 5: Use of sealing materials per area.

Knowledge of Portability of Water

Only 34% of households are reassured of the portability of the water, against 66% who are not at all reassured (Table 3).

Outside the area B, and to a lesser extent of the area A, where the majority of households seem confident about the portability of water tanks; in areas C, D and E households are not reassured of that portability (Figure 9).

Table 3: Results of the survey on knowledge of portability of water contained in water tanks

Questions/ Answers Survey area		Type of material used to seal?				Nature of the water stored in the tank?					Use of water?				Confidence on water potability?		
		Tiles	Food painting	Sika cement	Total	Rain water	SNDE* water	Drilling water	Water source	Total	Drinking water	Housework and cooking	Housework only	Total	Yes	No	Total
Area A	Number of tanks	19	0	5	24	1	18	3	2	24	7	14	3	24	14	10	24
	Percentage	7.4%	0%	2%	9.4%	0.4%	7%	1.2%	0.8%	9.4%	2.7%	5.5%	1.2%	9.4%	5.5%	3.9%	9.4%
Area B	Number of tanks	67	6	25	98	36	21	19	22	98	7	61	30	98	58	40	98
	Percentage	26.2%	2.3%	9.7%	38.3%	14.1%	8.2%	7.4%	8.6%	38.3%	2.7%	23.8%	11.7%	38.3%	22.7%	15.6%	38.3%
Area C	Number of tanks	46	1	5	52	14	18	19	1	52	10	37	5	52	7	45	52
	Percentage	18%	0.4%	2%	20.3%	5.5%	7%	7.4%	0.4%	20.3%	3.9%	14.4%	2%	20.3%	2.7%	17.6%	20.3%
Area D	Number of tanks	21	5	19	45	8	31	3	3	45	5	34	6	45	6	39	45
	Percentage	8.2%	2%	7.4%	17.6%	3.1%	12.1%	1.2%	1.2%	17.6%	2%	13.3%	2.3%	17.6%	2.3%	15.3%	17.6%
Area E	Number of tanks	11	1	25	37	9	6	1	21	37	3	32	2	37	2	35	37
	Percentage	4.3%	0.4%	9.7%	14.4%	3.5%	2.3%	0.4%	8.2%	14.4%	1.2%	12.5%	0.8%	14.4%	0.8%	13.6%	14.4%
Total all areas	Number of tanks	164	13	79	256	68	94	45	49	256	32	178	46	256	87	169	256
	Percentage	64.1%	5.1%	30.8%	100%	26.6%	36.7%	17.6%	19.1%	100%	12.5%	69.5%	18%	100%	34%	66%	100%

Table 4: Frequency of use of sealants for tanks in reinforced concrete and in masonry

Designation	Reinforced concrete	Masonry
Tiles	74(52%)	90(80%)
Food painting	4(3%)	9(8%)
Sika cement	65(45%)	14(12%)

Total	143(100%)	113(100%)

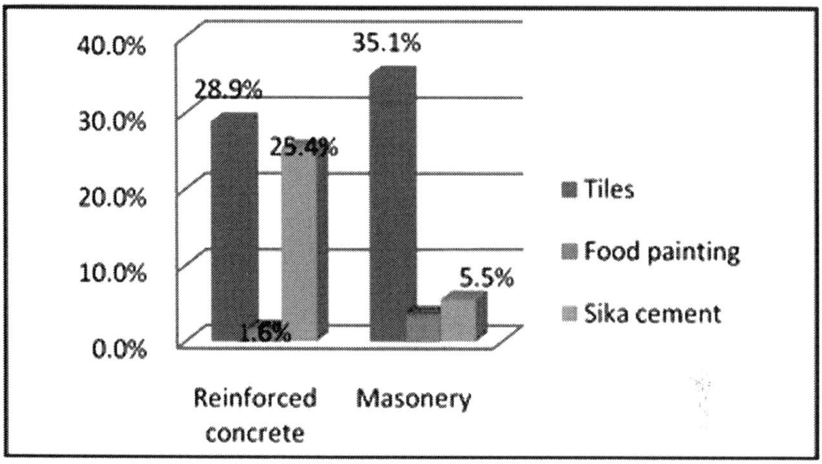

Figure 6: Use of sealing materials for reinforced concrete tanks and masonry tanks.

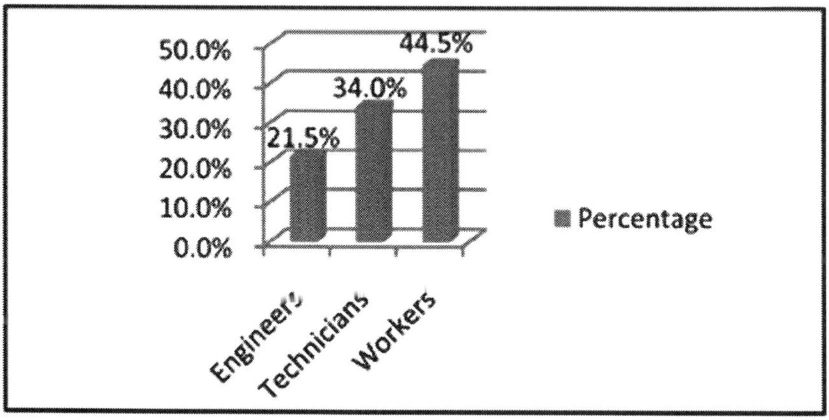

Figure 7: Qualification of trades for the construction of tanks.

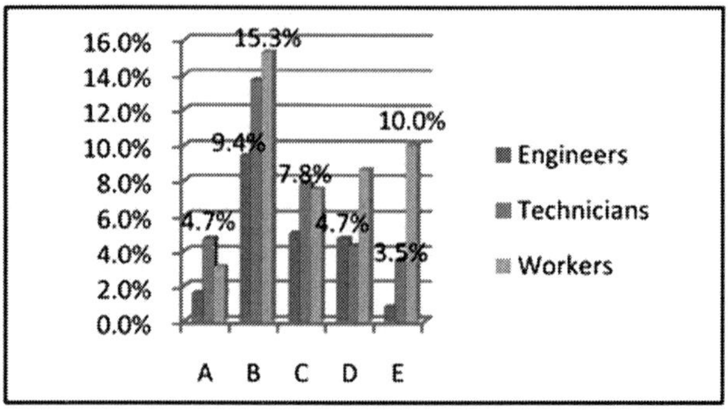

Figure 8: Distribution of trades per area.

Moreover, the largest number of households ignores the fact that environmental conditions can infect the water contained in the tanks (49.6%). However, those who are aware of the possible contamination of the water by the surrounding external environment are still quite numerous; they represent 46.1% of the total (Table 5).

We see that in areas A, B and D, households recognize the possibility that the water is infected by the external environment surrounding the tanks (Figure 10).

On the other hand, in the areas C and D, households in majority seem to do not worry about the possible contamination of the water stored in tanks.

Use of Water Tanks

The water stored in tanks is preferably used for the housework and cooking for over 69.5% of owners (Table 3). Those who use it only for housework represent 18% of the whole households. The stored water is used as drinking water by only 12.5% of the proprietors.

It is remarkable that in all areas, households mainly use water stored in tanks for housework and cooking (Figure 11). However, some owners also use it as drinking water and this in all the areas selected for the study.

Regarding the storage duration of the water in the tanks, it appears that the largest number of households store water for 3 to 4 weeks. They represent 52.4% of the total (Table 5). The percentage of those who store water for a very long time, more than five weeks, is still high (32%). Only 12% of households conserve water for a short period (2 weeks). In Figure 12 we have represented the water storage duration in tanks per area.

In all areas except area A, the majority of households store water in tanks for 3 to 4 weeks, thus confirming the general trend.

Discussion

The construction of underground water tanks in reinforced concrete requires the sealing problem is properly solved. In fact, concrete is a porous material by nature and prone to cracking. These two weaknesses led to consider layers or special treatments. The use of a coating of earthenware tiles, which is apparently the most appropriate solution, still leaves many users to be skeptical on the portability of the water stored in tanks. The water tightness of the tanks could be damaged in two types:

- The stripping of the coating;
- Cracks, classified into three types of categories: accidental cracks, withdrawal cracks, cracks under normal operating stresses.

Table 5: Results of the survey on the knowledge of the water storage duration in tanks

Questions/Answers Investigation area		Justification of water potability?			Water Storage duration in the tank?				Can the natural environmental conditions infect the water stored in the tanks?			
		Justified	Ujustified	Total	2 weeks	3 to 4 weeks	More than 5 weeks	Total	Yes	No	Without any opinion	Total
[illegible]	Number of tanks	14	10	24	7	4	13	24	19	3	2	24
	Percentage	5.5%	3.9%	9.4%	2.7%	1.6%	5.1%	9.4%	7.4%	1.2%	0.8%	9.4%
Area B	Number of tanks	98	0	98	9	52	37	98	58	40	0	98
	Percentage	38.3%	0%	38.3%	3.5%	20.3%	14.4%	38.3%	22.7%	15.6%	0%	38.3%
Area C	Number of tanks	7	45	52	19	27	6	52	16	27	9	52
	Percentage	2.7%	17.6%	20.3%	7.4%	10.6%	2.3%	20.3%	6.3%	10.5%	3.5%	20.3%
Area D	Number of tanks	6	39	45	2	30	13	45	23	22	0	45
	Percentage	2.3%	15.2%	17.6%	0.8%	11.7%	5.1%	17.6%	9%	8.6%	0%	17.6%
Area E	Number of tanks	2	35	37	3	21	13	37	2	35	0	37
	Percentage	0.8%	13.6%	14.4%	1.2%	8.2%	5.1%	14.4%	0.8%	13.6%	0%	14.4%
Total all areas	Number of tanks	127	129	256	40	134	82	256	118	127	11	256
	Percentage	49.6%	50.4%	100%	15.6%	52.4%	32%	100%	46.1%	49.6%	4.3%	100%

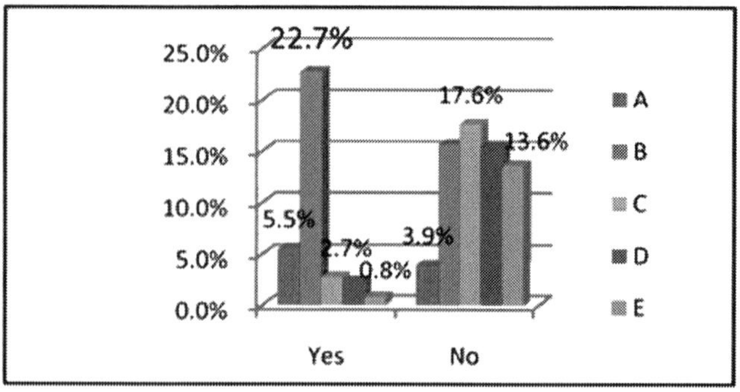

Figure 9: Confidence on the portability of water tank per area.

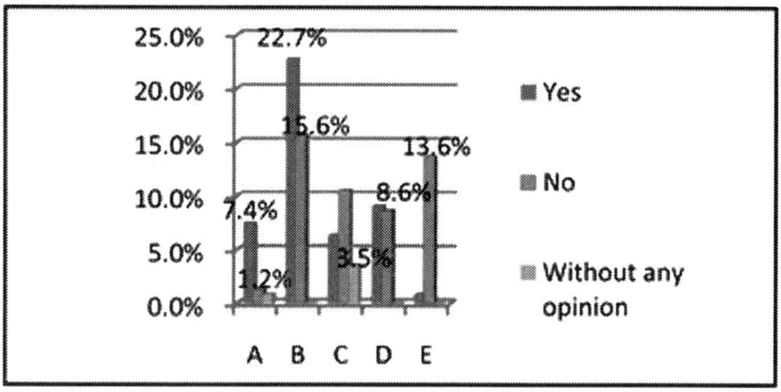

Figure 10: Knowledge of the possibility of water tank infection.

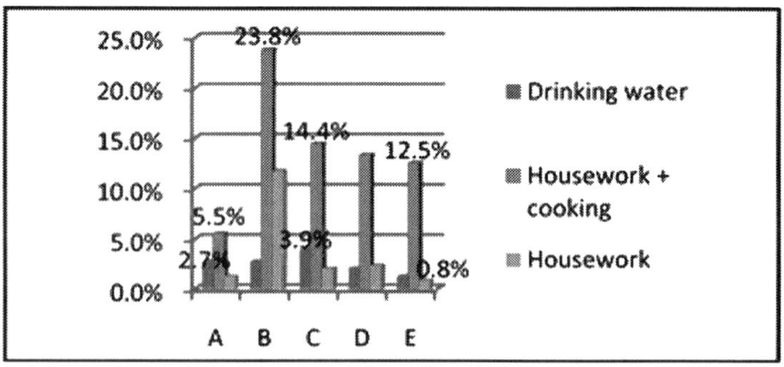

Figure 11: Use of water stored in tanks per area.

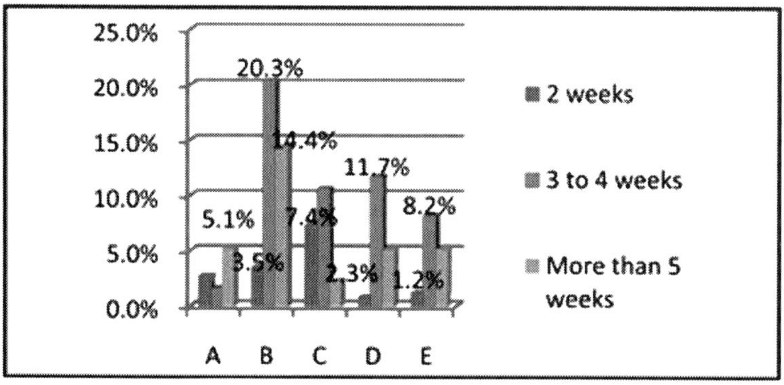

Figure 12: Water storage duration in tanks per area.

Stripping of the coating and the cracks are mainly due to the technological choice of materials, formulation, implementation and cure.

Infiltration of external elements that may contaminate the water stored in tanks are possible mainly through the cracks, as it is recognized that the concrete itself could be considered watertight [7]. We understand why for tanks constructed in masonry, more vulnerable to infiltration, sealing is provided almost exclusively with earthenware tiles (35.1% of tanks), compared to Sikacement (5.5%) and painting food (3.5%).

The lack of scientific publications touching on the type of studied topic in this work does not allow a comparison with any previous results, as we would wish.

CONCLUSIONS

On the whole, reinforced concrete tanks (56%) are preferred to those that constructed in masonry. The workers are the trade that carry out the most of work (44.5% of tanks), far ahead of the engineers (21.5% of tanks) and technicians (34% of the tanks). This important disqualification could have a direct impact on the quality of tanks built. For sealing the earthenware tiles are the most used with about 64% of the water tanks sealed. Despite these precautions to make the walls of the tanks to be tight, 66% of households suspect the portability of water stored. A significant number of households (46%) fear a possible contamination of the stored water by infiltration of noxious substances through the walls of the tanks. So, the stored water is it used by the immense majority of households (88%) only for cooking and/or housework. There are only 12% of owners who use it as drinking water.

It would be interesting to examine the hydro-chemical parameters of the water stored in underground tanks, which would be a way to prove the infiltration of outside elements, especially in areas where the soil is very wet. Given the difficulty of ensuring a hermetic sealing of reinforced concrete tanks, it could be recommended to install the tanks height rather than bury them.

REFERENCES

1. Vaishnav, M.M. and Dewangan, S. (2012) Analytical and Statistical Evaluation of Surface and Sub-Surface Water of BALCO Industrial Area, Korba, C. G., India. International Journal of Environmental Sciences, 2, 1369-1379.http://dx.doi.org/10.6088/ijes.00202030023.
2. Matini, L., Moutou, J.M. and Kongo-Mantono, M.S. (2009) Evaluation Hydro-Chimique des Eaux Souterraines en Milieu Urbain au Sud-Ouest de Brazzaville, Congo. Afrique Science, 5, 82-98.
3. Mahmood, A., Muqbool, W. and Mumtaz, M.W. (2011) Application of Multivariate Statistical Techniques for the Characterization of Ground Water Quality of Lahore, Gujranwala and Sialkot (Pakistan). Pakistan Journal of Analytical & Environmental Chemistry, 12, 102-112.
4. Liu, C.-W., Lin, K.-H. and Kuo, Y.-M. (2003) Application of Factor Analysis in the Assessment of Groundwater Quality in a Blackfoot Disease Area in Taiwan. The Science of the Total Environment, Elsevier, 77-89. http://dx.doi.org/10.1016/S0048-9697(02)00683-6.
5. Sharaky, A.M., Atta, S.A., El Hassanein, A.S. and Khallal, K.M.A. (2007) Hydrogeochemistry of Groundwater in the Western Nile Delta Aquifers, Egypt. 2nd International Conference of the Geology of Tethys, Cairo, 19-21 March, 29 p.

6. Dalton, M.G. and Upchurch, S.B. (1978) Interpretation of Hydrochemical Facies by Factor Analysis. Ground Water, 16, 228-233. http://dx.doi.org/10.1111/j.1745-6584.1978.tb03229.x.

Chapter 7

Achievements of Truss Models for Reinforced Concrete Structures

Panagis G. Papadopoulos, Hariton Xenidis, Panos Lazaridis, Andreas Diamantopoulos, Periklis Lambrou, and Yannis Arethas

Department of Civil Engineering, Aristotle University of Thessaloniki, Thessaloniki, Greece

ABSTRACT

Achievements are presented for truss models of RC structures developed in previous years: 1) Two constitutive models, biaxial and triaxial, are based on regular trusses, with bars obeying nonlinear uniaxial **σ-ε** laws of material under simulation; both models have been compared with test results and show a dependence of Poisson ratio on curvature of **σ-ε** law; 2) A truss finite element has been used in the nonlinear static and dynamic analysis of plane RC frames; it has been compared with test results and describes, in a simple way, the formation of plastic hinges; 3) Thanks to the very simple geometry of a truss, the equilibrium equations can be easily written and the stiffness matrix can be easily updated, both with respect to the deformed truss, within each step of a static incremental loading or within each time step of a dynamic analysis, so that to take into account geometric nonlinearities. So the confinement of a RC column is interpreted as a structural stability effect of concrete. And a significant role of the transverse reinforcement is revealed, that of preventing, by its close spacing and sufficient amount, the buckling of inner longitudinal concrete struts, which would lead to a global instability of the RC column; 4) The proposed truss model is statically indeterminate, so it exhibits some features, which are not met by the "strut-and-tie" model.

INTRODUCTION

In 1967, in a pioneering work [1], D. Ngo and A. C. Scordelis presented a detailed finite element model for a RC beam, in which separate finite elements are used for concrete and steel reinforcement. The material nonlinearities of the reinforcement can be easily described by the nonlinear uniaxial σ-ε law of a bar element. However, it is difficult to represent the nonlinear biaxial or triaxial stress-strain behavior of concrete or any other material. The relevant problems are discussed in two state-of-theart reports on nonlinear finite element analysis of RC structures, one by P. G. Bergan and I. Holand in 1979 [2] and another in a special publication of ASCE in 1982 [3], written by specialists on this field, under the co-ordination of A. C. Scordelis. Also, the difficulties appearing in the application of finite elements to nonlinear problems have been discussed in the series of three Conferences F. E. No. Mech. (Finite Elements in Nonlinear Mechanics), organized by J. H. Argyris in the Institute of Statics and Dynamics, University of Stuttgart, Germany in the years 1978, 1981, 1984 [4].

In order to describe the nonlinear biaxial or triaxial stress-strain behavior of a structure by the Finite Element Method, constitutive models for the structural materials have to be developed in order to be embodied in the individual finite elements. Efforts to develop such constitutive models have been made by many researchers, e.g. plasticity models by W. F. Chen [5] and Z. Mroz [6], the plastic-fracturing model of Z. P. Bazant [7], as well as the more practical contributions by D. Darwin [8] and K. I. Willam [9], for nonlinear, biaxial and triaxial, respectively, stress-strain behavior of concrete.

In 1977 [10], N. J. Burt and J. W. Dougill presented a random network constitutive model, in order to describe the nonlinear biaxial stress-strain law of a material, and noticed that equivalent results can be obtained by use of simple regular networks. By applying this idea, P. G. Papadopoulos developed in 1984 and 1986 [11,12] a biaxial and a triaxial network constitutive model, based on a regular plane octagon and a regular space rhombic dodecahedron, respectively, in which sides and diagonals are bars obeying the nonlinear uniaxial σ-ε laws of the material under simulation. Results from the above network constitutive models have been found in satisfactory approximation with corresponding published test results [13-15].

Trusses have been used not only in constitutive models, but also in finite elements of structures. In 1978 [16], E. Absi, in his "theorie des equivalences" stated that simple truss finite elements give equivalent results with the usual more complicated continuum finite elements. This idea was extended to problems with material nonlinearities and to the nonlinear static and dynamic analysis of plane RC frames by P. G. Papadopoulos [17,18]. A simple truss finite element was proposed, based on a plane rectangle in which all sides and diagonals are bars obeying nonlinear uniaxial σ-ε laws of concrete or steel. So, the nonlinear biaxial stress-strain behavior of the element is, in a simple way, described, thus the embodying of a constitutive law in the individual finite elements is no more needed. Results from nonlinear static analysis for cyclic loading, as well as nonlinear seismic dynamic analysis of simple plane RC frames, by the proposed truss RC element, were compared with corresponding published test results and found in a satisfactory approximation with them [19,20]. As the bars of the proposed finite element include the main material nonlinearities of concrete and steel, that is concrete tensile cracking and ultimate compressive strength, as well as tensile yield of reinforcement, the proposed truss model can, in a simple way, describes the formation of plastic hinges in a RC frame.

Afterwards, some other versions of E. Absi ideas for truss finite elements were developed for plane structures, under various names but all similar to each other, e.g. "truss analogy" in 1997 [21] for steel structures, "lattice model" in 1997 [22] and "lumped stress model" in 2002 [23], the latter two for RC structures.

In 1987 [24] J. Schlaich invented the so called "strutand-tie" model, which is a statically determinate truss model, consisting of concrete and steel bars. These bars include the main material nonlinearities of a RC structure. So, the "strut-and-tie" model can effectively describe the main stress-strain states of a RC structure, that is bending, shear and even torsion in 3D, thus it has been proved as a very useful practical tool in analysis and design of RC structures.

The "strut-and-tie" model has been further developed by other researchers, as by T. T. Hsu in 1993 [25], by F. J. Vecchio and M. P. Collins in 1993 [26], as well as by ASCE-ACI Committee on shear and torsion in 1998 [27].

The proposed here truss model is a statically indeterminate structure, so it exhibits some features that are not met by the statically determinate "strut-and-tie" model:

1) It can describe lateral expansion (Poisson ratio) effect.

2) It takes into account geometric nonlinearities, by writing the equilibrium equations and updating the stiffness matrix, both with respect to the deformed truss, within each step of a static incremental loading or within each time step of a dynamic analysis. This is easily achieved thanks to the very simple geometry of a truss.

By this proposed truss model which includes geometric nonlinearities, the confinement of a RC column is interpreted as a structural stability effect of concrete [28-30].

And, beyond the already known roles of the transverse reinforcement [31-33] (that is, shear transfer, reduction of concrete spalling, preventing of buckling of longitudinal reinforcement, increase of compressive stiffness, strength and ductility of the confined concrete core), another significant role of the transverse reinforcement is revealed by the proposed truss model with structural instability, that of retarding and even preventing, by its close spacing and sufficient amount, the buckling of inner longitudinal concrete struts, which would lead to a global instability of the RC column.

Results from the application of this proposed model with structural instability on RC column confinement have been found in a satisfactory approximation to Codes requirements [34-36], regarding the spacing and amount of transverse reinforcement, which, in turn, are based on test results, too.

In the following, some of the achievements of the above proposed truss models for nonlinear analysis of structures, mainly RC structures, proposed in previous years, will be described in more detail.

TRUSS CONSTITUTIVE MODELS

A biaxial and a triaxial constitutive model for the nonlinear stress-strain law of a material have been developed [11,12], based on a regular plane octagon and a regular space rhombic dodecahedron, respectively, in which sides and diagonals are bars obeying the nonlinear uniaxial σ-ε law of the material under simulation. So, in a simple way, by the nonlinear uniaxial σ-ε laws of the bars, the nonlinear biaxial or triaxial stress-strain behavior of the whole truss is described. Results from the above truss constitutive models, for various loading histories, have been found in satisfactory approximation to corresponding published test results [13-15]. Both above truss constitutive models show a dependence of the Poisson ratio value ν on the curvature κ of the nonlinear uniaxial σ-ε law of the material under consideration, as shown in Figure 1.

TRUSS FINITE ELEMENT FOR PLANE RC FRAME

A truss finite element is proposed for beams of a plane RC frame, based on a rectangle, in which all sides and diagonals are bars obeying nonlinear uniaxial σ-ε laws of concrete or steel, as shown in Figure 2. The σ-ε law of concrete bars includes tensile cracking, ultimate compressive strength, as well as loading-unloading rules after compressive yield. Whereas, the σ-ε law of a steel bar includes ultimate tensile and compressive strengths, as well as loading-unloading rules after tensile or compressive yielding.

DETERMINATION OF BAR SECTIONS

In the above proposed truss finite element for beams of plane RC frames, the cross-section areas of steel bars are reasonably and easily determined as sums of sections of the corresponding steel reinforcing bars. Whereas, in order to determine the cross-sections areas A_1, A_2, A_3 of the concrete bars of the truss element as shown in Figure 3(b), we have to compare it to the corresponding continuum concrete beam element of Figure 3(a), as regards three representative stress-strain states in the linear elastic region. And we chose, as such characteristic states, the pure bending, the confined axial deformation as well as the confined transverse deformation.

For the pure bending shown in Figure 3(c), the curvature angle of the beam element is $\Delta\varphi = Ml / EJ$ where $J=wh^3/12$, whereas for the truss element $\Delta\varphi = 2\Delta l / h$

Where $\Delta l = \frac{M}{h} l / A_1$ By combining the above equations we obtain $A_1=wh/6$.

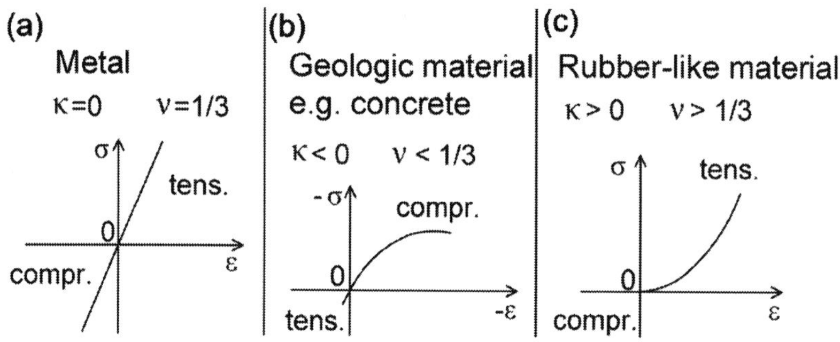

Figure 1: Dependence of Poisson ratio ν on the curvature κ of nonlinear uniaxial σ-ε law of the material. (a) Metal $\kappa = 0$ g $\nu = 1/3$; (b) Geologic material e.g. concrete $\kappa < 0$ g $\nu < 1/3$; (c) Rubber-like material $\kappa > 0$ g $\nu > 1/3$.

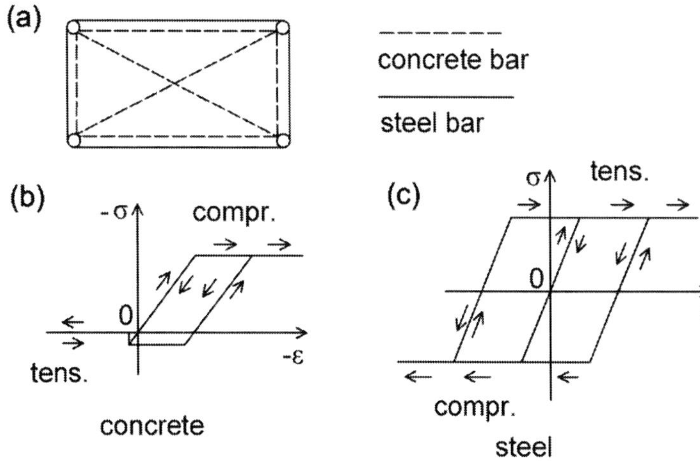

Figure 2: (a) Truss finite element for beam of a plane RC frame, with concrete and steel bars; (b) Nonlinear uniaxial σ-ε law of concrete bars; (c) Nonlinear uniaxial σ-ε law of steel bars.

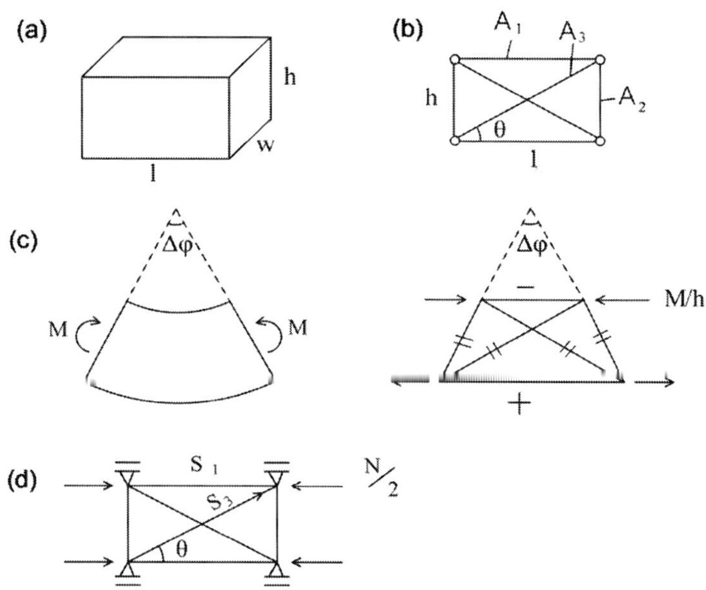

Figure 3: Comparison between characteristic stress-strain states of the concrete beam element and the corresponding truss element in order to deter-

mine the concrete bar sections. (a) Concrete beam element; (b) Corresponding truss element; (c) Pure bending; (d) Confined axial compression.

For the confined axial deformation, the elasticity theory gives

$$\sigma_x = \frac{E}{1-v^2}\varepsilon_{x.}$$

For $v \approx 0.2, 1-v^2 \approx 1$, thus $\sigma_x = N/wh$ and $\varepsilon_x = \Delta l/l.$, in the corresponding state of the truss element shown in Figure 3(d), we have $S_1 + S_3\cos\theta = N/2$

Where $S_1 = \frac{EA_1}{1}\Delta 1$ and $S_3 = \frac{EA_3}{1/\cos\theta}\Delta 1\cos\theta.$
From combination of above equations, we obtain

$$A_3 = wh/3\cos^3\theta$$

From similar considerations for confined transverse deformation, we obtain

$$A_2 = wl/2 - (wh/3)3tg^3\theta\,.$$

Obviously, when the angle θ tends to zero, $\theta \to 0$, the sections tend to $A_3 \to wh/3$ and $A_2 \to w1/2$

NONLINEAR STATIC ANALYSIS

The incremental loading of the structure is preferably performed by strain control, which is a more stable procedure than stress control. The material nonlinearities are taken into account by the variations of the elasticity moduli E of the bars during the loading. Whereas, in order to take into account geometric nonlinearities, the equilibrium equations are written and the global stiffness matrix updated, both with respect to the deformed truss, within each step of incremental loading. The local stiffness matrix of a bar in 2D, consisting of elastic and geometric part, is:

$$\mathbf{k} = \mathbf{k}_E + \mathbf{k}_G = \frac{EA}{l_o}\begin{pmatrix} c_x^2 & c_xc_y \\ c_xc_y & c_y^2 \end{pmatrix} + \frac{N}{l}\begin{pmatrix} c_y^2 & -c_xc_y \\ -c_xc_y & c_x^2 \end{pmatrix}$$

where A section, l_o undeformed length, l present length, N axial force and c_x, c_y direction cosines of the bar.

Whereas, the global stiffness matrix of the truss is:

$$\mathbf{K} = \mathbf{B}\mathrm{diag}\left(\mathbf{k}_i\right)\mathbf{B}^t \quad i = 1 \cdots n_b$$

where B Boolean linkage matrix and n_b number of bars of the truss.

Based on the proposed algorithm, a very short computer program, with only about 200 FORTRAN instructions, has been developed, for the nonlinear static analysis of a truss model of a plane RC frame.

NONLINEAR DYNAMIC ANALYSIS

A lumped mass is assigned to every free node of the truss. Zero damping and zero initial velocities are assumed. The resulting initial value problem:

$$\dot{\mathbf{y}} = \mathbf{q}\left(t, \mathbf{y}\right), \ \mathbf{y}\left(0\right) = \mathbf{y}_0 ,$$

Where the state vector is y={r,v,c} with r, v positions and velocities of nodes and c constitutive variables of the bars, is solved by the step-by-step algorithm of trapezoidal rule, which coincides with the Newmark's algorithm of constant average acceleration:

$$\mathbf{y}_{n+1} = \mathbf{y}_n + \frac{1}{2}\left[\mathbf{q}\left(t_n, \mathbf{y}_n\right) + \mathbf{q}\left(t_{n+1}, \mathbf{y}_{n+1}\right)\right]\Delta t ,$$

Combined with a predictor-corrector technique with two corrections per step, $PE(CE)^2$ [37]. So, there is no need to solve an algebraic system within each step of the algorithm.

The stability criterion of the algorithm is $\omega_{max}\Delta t < 2.0$ rad and the accuracy criterion is $\omega_{max}\Delta t < 0.5$ rad, that is $\Delta t < T_{min} / 4\pi$, which dictates the choosing of the time step-length Δt of the algorithm.

An upper bound for the normal frequencies can be found from the norm of the matrix $M^{-1}K$, where M mass matrix and K stiffness matrix of the structure:

$$\omega_{\max} < \left\| \mathbf{M}^{-1}\mathbf{K} \right\|$$

Based on the proposed algorithm, a very short computer program has been developed, with only about 150 FORTRAN instructions, for the nonlinear dynamic analysis of a truss model of a RC frame.

APPLICATIONS TO ANALYSIS OF SIMPLE PLANE RC FRAMES

The above proposed truss finite element for plane RC frames, as well as the proposed algorithms for nonlinear static and dynamic analysis, have been applied to the nonlinear static analysis of a simple plane RC frame for cyclic loading [17], as well as to the nonlinear dynamic seismic analysis of a simple plane RC frame [18]. The results of these analyses have been found in satisfactory approximation with corresponding published test results [19, 20].

As the nonlinear uniaxial σ-ε laws, of the bars of the proposed truss model, include all the main material nonlinearities of a RC structure, that is tensile cracking and ultimate compressive strength of concrete, as well as tensile yielding of steel reinforcement, the formation of plastic hinges in a RC frame is, in a simple way, described, as shown in Figure 4.

APPLICATION TO CONFINEMENT OF A RC COLUMN

In order to take into account geometric nonlinearities, by the proposed truss model, the equilibrium equations are written and the stiffness matrix is updated, both with respect to the deformed truss, within each step of a static incremental loading or within each time step of a dynamic analysis. This is easily achieved thanks to the very simple geometry of a truss.

As the proposed truss model includes geometric nonlinearities, it interprets the confinement of a RC column as a structural stability effect of concrete [28-30].

In Figure 5(a), the compressive axial σ-ε diagram of a confined RC column is shown. An early small drop of the stress $-\Delta\sigma$ is observed, which is due to a local instability because of spalling (buckling) of outer concrete. As the loading further increases, preferably by strain control, for a significant value of the compressive axial deformation, the stress σ suddenly drops to zero, which is an obvious mark of global structural instability, observed in experiments and verified by the proposed truss model, too.

In Figure 5(b), a part of a confined RC column, between two successive sets of transverse reinforcement, is shown. The longitudinal reinforcement is omitted, for simplicity. As the compressive axial loading N gradually increases, a lateral expansion of concrete occurs. For a significant compressive axial deformation, because of the large lateral expansion of concrete, a tensile yielding of the transverse reinforcement occurs, which implies a further lateral expansion of concrete. So, wide longitudinal-vertical concrete cracks are formed, and between such successive concrete cracks, inner longitudinal-vertical concrete struts are formed, which tend to buckle, leading to a global instability of the RC column.

Beyond the already known roles of transverse reinforcement in a RC column [31-33] (that is shear transfer, reduction of concrete spalling, preventing of buckling of longitudinal reinforcement, increase of compressive axial stiffness, strength and ductility of the confined concrete core), another significant role of the transverse reinforcement is revealed by the proposed truss model with structural instability, that of retarding and even preventing, by its close spacing s and sufficient amount ρ (mechanical ratio), the buckling of the above inner longitudinal-vertical concrete struts, which would lead to a global instability of the RC column.

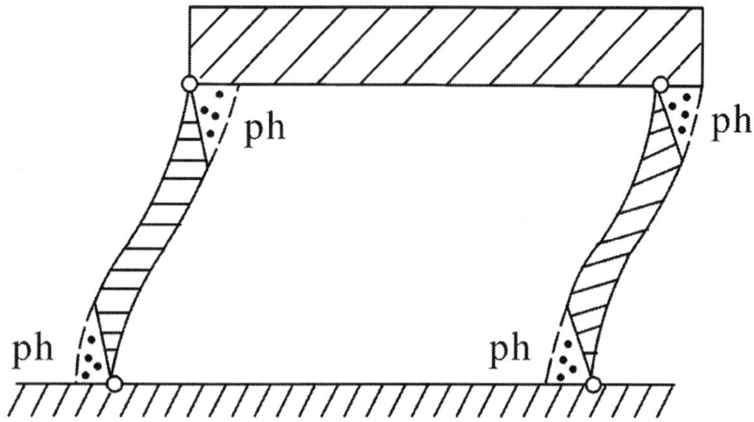

Figure 4: Description of formation of plastic hinges, in a RC frame, by the proposed truss model. "……" cracked concrete. "------" reinforcement in tensile yielding. "//////" rigid parts. "ph" plastic hinges.

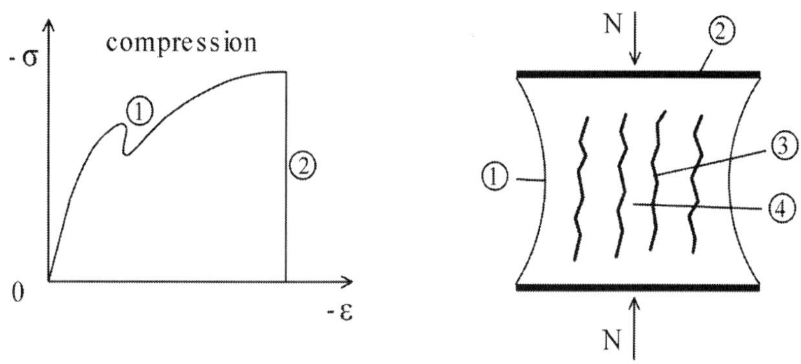

Figure 5: (a) Compressive axial σ-ε diagram of a confined RC column. 1. Early small drop of stress due to spalling of outer concrete. 2. For a significant value of compressive axial deformation, the stress σ suddenly drops to zero, which is a mark of global structural instability; (b) A part of a confined RC column between two successive sets of transverse reinforcement. 1. Spalling of outer concrete. 2. Transverse reinforcement in tensile yielding. 3. Longitudinal concrete cracks. 4. Longitudinal concrete struts.

Results, from application of the proposed truss model with structural instability to the confinement of RC columns, have been found in satisfactory approximation with corresponding requirements of codes [34-36], regarding the spacing s and the mechanical ratio of transverse reinforcement; these requirements are, in turn, based on test results, too.

CONCLUSIONS

Some achievements have been presented for truss models of structures, mainly RC structures, which have been developed in previous years and found in satisfactory approximation with test results and Codes requirements:

- N. J. Burt and J. W. Dougill developed in 1977 [10] random network constitutive models and stated that equivalent results can be, in a simple way, obtained by regular networks. This idea was realized in 1984 [11] and 1986 [12] by two network constitutive models, a biaxial and triaxial one, based on a regular plane octagon and on a regular space rhombic dodecahedron, respectively, in which sides and diagonals are bars obeying the nonlinear uniaxial σ-ε law of the material under simulation. Both models show a dependence of Poisson ratio on the curvature of the nonlinear uniaxial σ-ε law of the material.
- E. Absi in 1978 [16], in his "theorie des equivalences", stated that simple truss finite elements give equivalent results with the usual more complicated continuum finite elements. This idea was extended in 1988 [17, 18] to structures with material nonlinearities, and applied particularly to the nonlinear static and dynamic analysis of plane RC frames. As the individual bars of the proposed truss finite element include, in their uniaxial σ-ε laws, the main material nonlinearities of a RC structure, that is the concrete tensile cracking, the reinforcement tensile yield, as well as the ultimate compressive strength of concrete, the formation of plastic hinges in a RC frame can be, in a simple way, described.
- Compared to the "strut-and-tie" model for RC structures, invented by J.Schlaich in 1987 [24] and further developed by other researchers, which proved as a very effective tool in the analysis of RC structures, the proposed here truss model exhibits the difference that it is a statically indeterminate, whereas the "strut-and-tie" model is statically determinate. So, the proposed truss model has some features that are not met by the "strut-

and-tie" model: a) It can describe lateral expansion (Poisson ratio) effect. b) It takes into account geometric nonlinearities, by writing equilibrium equations and by updating stiffness matrix, both with respect to the deformed truss, within each step of a nonlinear static or dynamic analysis. So, it interpreted in 1999 [31] the confinement of a RC column as a structural stability effect of concrete. And revealed a significant role of transverse reinforcement, that of retarding and even preventing, by its close spacing and sufficient amount, the buckling of inner longitudinal concrete struts, which would lead to a global instability of the RC column.

REFERENCES

1. D. Ngo and A. C. Scordelis, "Finite Element Analysis of Reinforced Concrete Beams," ACI Journal, Vol. 64, 1967, pp. 152-163.
2. P. G. Bergan and I. Holand, "Nonlinear Finite Element Analysis of Concrete Structures," Computer Methods in Applied Mechanics and Engineering, Vol. 17-18, 1979, pp. 443-467. doi:10.1016/0045-7825(79)90027-6
3. A. C. Scordelis, Editor, ASCE Task Committee on Concrete and Masonry Structures, "State-of-the-Art Report on Finite Element Analysis of Reinforced Concrete," ASCE Special Publication, 1982.
4. J. H. Argyris, Organizer, International Conferences F.E.No.Mech. (Finite Elements in Nonlinear Mechanics). Institute for Statics and Dynamics, University of Stutgart, Germany, I.30 August-1 September 1978, II. 25-28 August 1981. III. 10-13 September 1984.
5. W. F. Chen and E. C. Ting. "Constitutive Models for Concrete Structures," Journal of Engineering Mechanics Division ASCE, Vol. 106, No. 1, 1980, pp. 1-19.
6. Z. Mroz, V. A. Norris and O. C. Zienkiewicz, "Application of an Anisotropic Hardening Model in the Analysis of Elastic-Plastic Deformation of Soils," Geotechnique, Vol. 29, 1979, pp. 1-34. doi:10.1680/geot.1979.29.1.1

7. Z. P. Bazant and S. S. Kim, "Plastic-Fracturing Theory for Concrete," Journal of Engineering Mechanics Division ASCE, Vol. 105, No. 3, 1979, pp. 407-428.
8. D. Darwin and D. A. Pecknold, "Analysis of Cyclic Loading of RC Structures," Computers and Structures, Vol. 7, No. 1, 1977, pp. 137-147. doi:10.1016/0045-7949(77)90068-2
9. K. J. Willam and E. P. Warnke, "Constitutive Model for the Triaxial Behavior of Concrete," Proceedings of IABSE, Structural Engineering Report 19, Section III, 1975, pp. 1-30.
10. N. J. Burt and J. W. Dougill, "Progressive Failure in a Model Heterogeneous Medium," Journal of Engineering Mechanics Division ASCE, Vol. 103, 1977, pp. 365-376.
11. P. G. Papadopoulos, "Biaxial Network Constitutive Model," Journal of Engineering Mechanics ASCE, Vol. 110, No. 3, 1984, pp. 449-464. doi:10.1061/(ASCE)0733-9399(1984)110:3(449)
12. P. G. Papadopoulos, "A Triaxial Network Constitutive Model," Computers and Structures, Vol. 23, 1986, pp. 497-501. doi:10.1016/0045-7949(86)90093-3
13. H. B. Kupfer, H. D. Hilsdorf and H. Rusch, "Behavior of Concrete under Biaxial Stresses," ACI Journal, Vol. 66, No. 8, 1969, pp. 656-666.
14. R. Palaniswamy and S. P. Shah, "Fracture and StressStrain Relationships of Concrete under Triaxial Compression," Journal of Structural Division ASCE, Vol. 100, 1974, pp. 901-916.
15. R. Scavuzzo, T. Stankowski, K. Gerstle and H.-Y. Ko, "Stress-Strain Curves for Concrete under Multiaxial Load Histories," University of Colorado, Boulder, 1983.
16. E. Absi, "Méthodes des Calcus Numerique en Elasticité," Eyrolles, Paris, 1978.
17. P. G. Papadopoulos, "Nonlinear Static Analysis of Reinforced Concrete Frames by Network Models," Advances in Engineering Software, Vol. 110, No. 3, 1988, pp. 114- 122.doi:10.1016/0141-1195(88)90010-1
18. P. G. Papadopoulos and C. G. Karayannis, "Seismic Analysis of R/C Frames by Network Models," Computers and Structures, Vol. 28, No. 4, 1988, pp. 481-494.doi:10.1016/0045-7949(88)90022-3

19. K. Stylianidis and G. Penelis, "Experimental Study of, bare and Infilled by Wall, One Story Frames under Cyclic shear Loading," 7th Greek Conference on Concrete, Vol. 2, Patra, 1985, pp. 47-55.

20. P. Hidalgo and R. W. Clough, "Earthquake Simulator Study of a Reinforced Concrete Frame," EERC Report 74-13, University of California, Berkeley, 1974.

21. S. C. Goel, B. Stojadinovicz and K. H. Lee, "Truss Analogy for Steel Moment Connections," Engineering Journal, Second Quarter 1997, pp. 43-53.

22. E. Schlangen and E. J. Garboczi, "Fracture Simulations of Concrete Using Lattice Models: Computational Aspects," Engineering Fracture Mechanics, Vol. 57, No. 2-3, 1997, pp. 319-332. doi:10.1016/S0013-7944(97)00010-6

23. F. Fraternali, M. Angelilo and A. Fortunato, "A Lumped Stress Method for Plane Elastic Problems and the Discrete Continuum Approximation," International Journal of Solids and Structures, Vol. 39, 2002, pp. 6211-6240. doi:10.1016/S0020-7683(02)00472-9

24. J. Schlaich, K. Schäfer and M. Jennewein, "Towards a Consistent Design of Structural Concrete," PCI Journal Special Report, Vol. 32, No. 3, 1987, pp. 75-150.

25. T. T. C. Hsu, "Unified Theory of Reinforced Concrete," CRC Press, 1993.

26. F. J. Vecchio and M. P. Collins, "Compression Response of Cracked Reinforced Concrete," Journal of Structural Engineering ASCE, Vol. 113, 1993, pp. 3590-3610.doi:10.1061/(ASCE)0733-9445(1993)119:12(3590)

27. ASCE-ACI Committee 445 on Shear and Torsion, "Recent Approaches to Shear Design of Structural Concrete. State-of-the-Art Report," Journal of Structural Engineering ASCE, Vol. 119, No. 12, 1998, pp. 1375-1417.

28. P. G. Papadopoulos and H. C. Xenidis, "A Truss Model with Structural Instability for the Confinement of Concrete Columns," Journal of EEE (European Earthquake Engineering), Part 2, 1999, pp. 57-79.

29. P. G. Papadopoulos, H. Xenidis, C. Karayannis, A. Diamantopoulos and P. Lambrou, "Confinement of Concrete Column Interpreted

as a Structural Stability Effect," 6th GRACM (Greek Association of Computational Mechanics) Conference, Thessaloniki, 19-21 June 2008.

30. P. G. Papadopoulos, H. Xenidis, D. Plasatis, P. Kiousis and C. Karayannis, "Concrete Stability Achieved by Confinement in a RC Column," 12th International Conference on Civil, Structural and Environmental Engineering Computing, Coordinator B.H.V. Topping, Madeira, Portugal, 1-4 September 2009.
31. K. Park, M. J. N. Priestley and W. D. Gill, "Ductility of Square Confined Concrete Columns," Journal of Structural Division ASCE, Vol. 108, No. 4, 1982, pp. 929-950.
32. S. Watson, F. A. Zahn and R. Park, "Confining Reinforcement for Concrete Columns," Journal of Structural Engineering ASCE, Vol. 120, No. 6, 1984, pp. 1798- 1849.
33. J. B. Mander, M. J. N. Priestley and R. Park, "Theoretical Stress-Strain Model for Confined Concrete," Journal of Structural Engineering ASCE, Vol. 114, No. 8, 1988, pp. 1804-1826. doi:10.1061/(ASCE)0733-9445(1988)114:8(1804)
34. Uniform Building Code 2, "Structural Engineering Design Provisions," Chapter 19. Concrete, 19.2.1. Reinforced Concrete Structures Resisting Forces Induced by Earthquake Motions 19.2.14. Frame Members Subjected to Bending and Axial Load, 1994, pp. 237-239.
35. New Zealand Standards 3101, "Code of Practice for the Design of Concrete Structures," Chapter 17, Members Subjected to Flexure and Axial Loads, Additional Seismic Requirements, 1989.
36. Eurocode 8, "Earthquake Resistant Design of Structures," Part 1-3. General Rules and Rules for Buildings. 2, Specific Rules for Concrete Buildings. 2.8. Provisions for Columns, Brussels, 1993, pp. 35-46.
37. P. G. Papadopoulos, "A Simple Algorithm for the Nonlinear Dynamic Analysis of Networks," Computers and Structures, Vol. 18, No. 1, 1984, pp. 1-8. doi:10.1016/0045-7949(84)90074-9

Chapter 8

Shale Stone and Fly Ash Landfill Use in Landslide Hazardous Area in Şirnak City with Foam Concrete

Yıldırım İ. Tosun

Mining Engineering Department, Engineering Faculty, Şırnak University, Şırnak, Turkey

ABSTRACT

Şırnak City and the surrounding areas are on steeper slopes. There are sliding large land masses or rocks. Underground water and harsh climatic conditions contain high risk hazard areas in urban living site with higher population density. In order to eliminate landslides and related events, significant precautions should be taken. The mapping of landslide risk may ease to take precautions. Even the application of landfill rock may reduce water content of soil. In this research, fly ash and Mine Waste shale stone were used with low density foam concrete. Waste mixture at certain proportions decreased cement use. Shale stone as fine aggregate instead of fly ash in specific proportions improved mechanical strength and porosity. Hence landslide hazardous area could be safer for urban living.

INTRODUCTION

In the developing countries, according to the topography of land in the mountainous areas, improper use of land increases the possibility of developing landslide. During earth excavations and constructions the landslide hazardous areas may not be considered [1] - [3].

Generally landslide hazardous area can be described as "possible sliding masses which make an angle with the surface" [4] - [6].Slope stability and landslide problem dealing with the ground have been closely worked for many years by geotechnical engineers. For this reason, the deep disorder of land masses and slope stability, flood, are known as natural disasters such as hurricanes, similar to leading to serious loss of life and property [7] - [9]. With the increase of population growth there is a great need for the fields in long natural slopes. In these fields, people require some excavation, drainage precautions, different urban constructions, specific vegetation which can protect natural balance. However, the natural land mass stability can be concerned of the reasons such as heavy rains or earthquakes [10] - [13].

The civil engineers are much interested in the foundation issues regarding the earth soil, surface created by the nature, the slope geometry and design of the structures built on land [14] - [18].The technological development parallel to the housing needs the high embankment, dams, large and deep excavations along with stability problems. Each year, few flood and landslide damages may cause to heavy loss of life. They also cause to loss of the millions of pounds in the world. The landslide in Turkey is one of the most important geotechnical hazards. Foam concrete produces much resistive low density porous structure in landfill by use of fly ash [19] - [23].Fine aggregate with waste fly ash can improve strength in foam concrete [12] [24] - [27].

In this study, ırnak city and the surrounding area were studied by geological mapping at 1/1000 scale and the soil units. The properties of every unit soil were determined. Landslide hazardous area provisionally was concerned in the first region's landslide area. The soil properties worked promotes the construction distributions in the future in ırnak City, promoting winter tourism, one of the popular Southeastern Anatolian land. The importance of civil constructions under the threat of landslide intended to draw attention to the urbanization of the area.

In order to minimize the landslide hazard, geological and geotechnical analysis of the land slopes were needed to be studied. In this study, in the area of south Şırnak City (Figure 1) 2 km circled from the center of the slopes S2 was investigated as much critical hazardous area due to some mass sliding occurred in the district. Geotechnical properties soils of four different locations were determined for examining the stability analysis. This project was carried out foam concrete application for hazardous landslide area for urban use, which will open workspace and environment covering the foundations and stabilities of civil engineering construction in the field of 4 km^2area. The laboratory studies followed by the polar coordinate system using a field study with the topographic maps and cross-sections of four slopes were prepared. Landfills with foam concrete were applied on the research site.

GEOLOGY OF THE ŞIRNAK CITY PROVINCE

Southeastern Anatolian Region has certain tectonic structures and stratigraphy containing the fault system and especially tension cracks along the ground to the depths. Underground waters, geothermal energy fluids flow the deep formations [28]-[30].North-south direction across the region with the effect of compression of the earth's crust has been subjected to a stretching east-west direction, and the resulting tension cracks along the earth have risen from the olivine magma.

Alluvium, dissociated limestone in the recent study area of ırnak City (Figure 2) is distributed among the alluvial deposits of the neighborhood. There are the limestone gravels in the area and very well rounded pebbles associated with soil. Alluvium soil, marly shale occurs in wide urban living area of ırnak City at the south, which field as seen in Figure 1 extends to the study area. In field observations, the Miocene aged limestone and dissociated limestone were determined. Thickness of this formation is highly variable among 2 - 30 m. Decrease in the slope of the land shows that a relatively small outcrop occurs.

Figure 1: Satellite image of the study area and the slopes in Şirnak City.

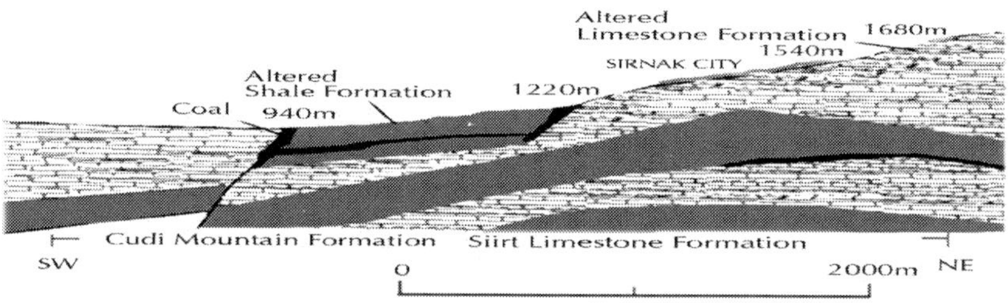

Figure 2: The geologic cross-section of the irnak City.

Southwestern limestone is moderately dissociated due to the deteriorated rock formations into the layers. Due to the resistance, large parts of rock at 50% - 90% rate are different. The cracks on the surface of the rocks are dominated by brown and blackish color. In the study area, the cross-section (Figure 2) shows the extent of moderately, or highly dissociated formation along the valley portion. In those regions, the separated segments of the shale are observed.

The study area to the north of the city contains fundamentally higher land slopes in the urban location. The observations of sorting and grading of seen alluvial fill show a thickness varies between 10 - 35 m. The high slope massive fills are concerned as active and potential hazardous landslide areas and are requiring for various examinations.

GEOTECHNICAL PROPERTIES

The representative contents of drilling logs were taken from the rectangular fields as seen from Figure 1. The representative samples of soil were taken from each different location over the slopes. The experiments conducted to determine the geotechnical characteristics regarding the Turkish Standards and American Standards (ASTM 3080) [31]-[33].Regarding the logs of wells of 58 m, 135 m, 60 m, 574 m wells, the mechanical properties of soils are given in Table 1.

In the study area of fine-grained portions of alluvium samples were taken from slopes as undisturbed and disturbed sample logs. The results of the experiments conducted on the samples of disturbed soil grain distribution curve, unit weight and consistency limit grain sizes are given in Table 1.

With the help of Shear box tests on undisturbed samples, the effective cohesion (c′) and effective shear resistance angle (φ°) belonging to the representative logs was found. TS standard and ASTM standards were used [33].During carrying out this experiment, the bulk density of the material, and the void ratio were determined with compression amount. Plastic and liquid limit results obtained in the experiments for each sample are given in Table 1. According to the classification of the soils in S1 and S2 landslide hazardous areas, the soils were determined as less plastic and non-plastic group, even the soils of S3 and S4 in the landslide hazardous area were determined as less plastic.

RESULTS AND DISCUSSION

Foam Concrete Landfill

Fly ash and Mine Waste Shale stone thrown as waste in Şırnak City of Turkey were used with foam concrete. Waste mixture at certain proportions of fly ash and shale stone were decreasing the use of cement. Fine shale aggregate in specific proportions were decreasing the use of fly ash. Certain proportion of fly ash and fine shale aggregate mixture used improved mechanical strength and porosity. Fine aggregate was reducing porosity rather than fly ash. Three different methods are performed in foam concrete production. Firstly, only foam concrete was used as binder of mixture fill. Secondly, foam concrete included fly ash. Thirdly, foam concrete included fly ash and fine shale stone. Effects of foam concrete on mechanical properties of fly ash and fine shale stone aggregate were studied intensively. Uniaxial and triaxial compression strengths of foam concrete landfills were tested in the laboratory ELE type press [26] [33].

Table 1: Properties soil formation in Şirnak City

Spec No.	**S1**	**S2**	**S3**	**S4**
Level (m)	1225	1221	1233	1227
Wopt, %	15.9	13.7	10.8	11.4
c′ (kPa)	1.2	4.7	5.3	2.5
□′	12.5	22.5	21	20
L_1(%)	26	15	28	17
P_1(%)	19	11	18	22
Ip (%)	10	9	8	12
γ_{sat}g/cm^3	2.4	2.5	2.4	2.3
γ_{ap}g/cm^3	1.65	1.6	1.78	1.6
γ_{dry}g/cm^3	1.82	1.76	1.9	1.7
γ_sg/cm^3	2.02	1.84	2.0	1.8
Spec. No.	S11	S22	S31	S41
γ_{max}g/cm^3	1.68	1.67	2.05	1.90
w_{opt}%	15.9	15.8	12.3	13.0
Permeability (k) (mD)	0.56	0.38	0.31	0.56

Another purpose of this study was to improve the ready-mixed foam concrete industry, and beneficiation from ırnak Coal Mine Waste shale stone and fly ash. The results contributed the widely efficient use of high amount of shale stone and fly ash, better workability of the foam concrete and the improved curing time. For this purpose, in the mixture of foam concrete fine shale aggregate was added 5%, 10%, 20% and 30% weight rates, fly ash was added 5%, 10%, 20% and 30% weight rates, respectively. Water/cement ratio were kept constant 3/1 - 4/1 volume rate. The prepared laboratory 10 × 10 × 10 cm cubic blocks was tested as the foam concrete mixture blocks. In conclusion, depending on the amount of the fly ash in foam concrete mixture, workability significantly improved and even significant the pressure increase was observed.

Compression strength values depending on the increasing amount of fly ash and foam concrete reduced. Increase in fly ash amount used in the foam concrete mixture by curing time was found to reduce the porosity.

The representative rock fill samples were taken from the construction foundation and strengths with the bulk density are shown in Table 2. For determination of rock types based on logging were carried out and the results showed poor and good rock classification as given in Tables 2-4.

In this study, coarse shale stone over 10 cm sizes as rock fill in the foundation and with foam concrete was used. The strength and the mechanical parameters of addition of rock fill at 10% weight rate as shale stone into the foam concrete were investigated. The results of those blocks are given in Table 3. The amount of fly ash on properties of water discharges in the landslide risk area were also investigated and briefly given in Table 4.

Procurement of Slopes with the Foam Concrete

Evaluation of the test results to determine the level of soil permeability regarding Table 1, S1, S2, S3 and S4 slopes may be critically hazardous on land constructions. It is observed that low permeable ground under construction as seen in Figure 3 caused sliding tension cracks and soil land slide.

Landfill of foam concrete performed was compared with the soil samples taken from the test results are given in Table 2. The pore pressure parameters foam concrete landfill at the optimum water content and maximum dry unit weight are given in Table 3. The determined values were used in the calculation of the stability of slopes. A natural slope does not affect the stability of the compression parameters. These parameters of the soil were improved by compacted landfill in the desired manner. Artificial compression parameters used directly on the slopes. If there is a risk of landslide hazard in a natural layer, compression case using these parameters are used in stability analysis. The use of anchorage or pile applications should be taken in consideration against to the any possible hazard of landslide in the front slope or the slope should be gradually compacted. In this study, an amount of the natural ground is excavated and compressed and foundation was filled by certain foam concrete and rock fill. The pore parameters were improved against to the hazard. In this case, the stability analysis of these improved parameters was used in the stability analysis of compressed ground.

Figure 3: S2 study area, construction work and sliding land soil and foam concrete landfill study area.

Table 2: The drilling data and calculated rqd by rmr rock classification parameters

Depth (m)	Drill Log (%)	RQD	RMR	Rock Quality
0.3 - 1.5	69.3 - 78.9	53.3 - 67.2	68	Good
1.5 - 3.0	58.4 - 90.2	50.6 - 78.5	63	Good
3.0 - 4.5	84.5 - 97.5	56.9 - 83.4	62	Good
4.5 - 6.0	96.1 - 94.3	44.2 - 90.9	47	Weak
6.0 - 7.5	94.6 - 96.9	43.9 - 45.3	47	Weak
7.5 - 9.0	15.5 - 25.0	48.1 - 43.7	44	Weak

9.0 - 10.5	16.4 - 22.0	49.4 - 46.2	44	Weak
10.5 - 12.0	96.3 - 98.2	47.4 - 45.3	44	Weak

Table 3: The mechanical parameter values of foam concrete landfill

Natural Unit Weight $_n$(g/cm^3)	1.21
Saturated Unit Weight $_d$(g/cm^3)	1.44
Cohesion, c (kPa)	135
Internal Friction Angle, (°)	34

Table 4: The parameters and the resulting strength of foam concrete landfill

Mixture in Foam Concrete	**σ, Strength MPa**	**Water Discharge (%)**	**γ_{ap} Unit Weight, g/cm^3**
Shale Stone, 5%	8.9 - 9.9	88.2	1.540
10%	12.4 - 13.9	85.5	1.620
20%	14.5 - 17.5	82.4	1.790
30%	19.1 - 20.3	77.7	1.850
Fly Ash, 5%	9.6 - 9.9	64.3	1.250
10%	11.5 - 12.5	62.7	1.380
20%	13.4 - 13.9	61.2	1.420
30%	14.3 - 14.9	60.3	1.580

c′ and φ′ values are taken as given in Table 3, respectively. In addition, the safety coefficient values were used as 1.3 - 1.5 in GEO5 program. The probable circular sliding surfaces were determined using the methods such as landslide, according to Fellenius, Bishop and Janbu calculations [34] - [36].

This kind of soil conditions in this study provide a circular or non-circular sliding close quarters to the top of the slope begin as a deep developing and ongoing planar surfaces. In these type of instabilities developed along the sliding surface in order to investigate the stability of the slope the Bishop method is commonly used [37] - [39] .

At the construction site foundation the foam concrete wells in 2 m diameters were excavated at elevation 1210 m to 10 m depth to sliding face. At creeping slopes hazard of mass slide may be prevented by rock fill. Even 1 meter depth foam concrete landfill horizontal columns constructed as seen in Figure 4.

With performed the foam concrete landfill application as constructed in Figure 4, hydraulic works at high rate water discharge through the foam concrete structure of landfill, even separately improved compression strength obtained by waste shale stone and fly ash use. Shale stone in the foam concrete mixture provided higher water discharges as given in Table 4. According to landfill construction seen from the Figure 5, the landfill length varies from 3 to 10 m horizontal length. Shale stone and fly ash concrete covered foam concrete tunnels and wells could easily provide the sliding face dry.

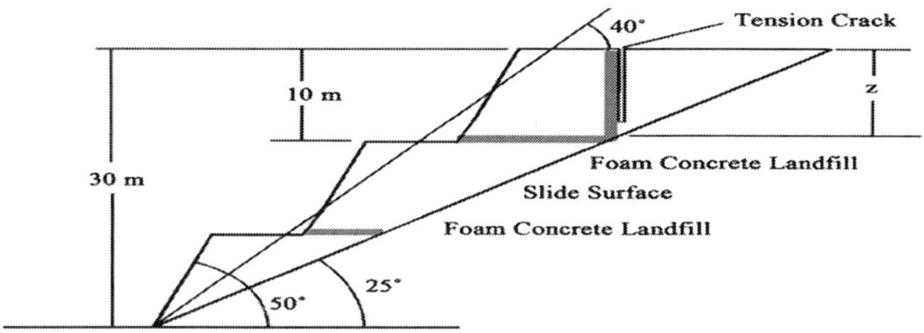

Figure 4: S2 Study area, foam concrete landfill application cross-section on hazardous landslide area.

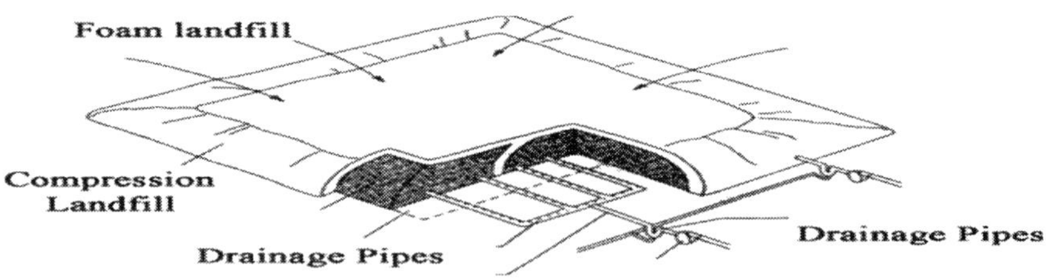

Figure 5: Construction foam concrete landfill application on hazardous landslide area.

The landslide S2 maximum elevation difference between the top and the heel point 35 m, 30 m maximum height of the slope divided three faces by excavation, surface slope angle is 43°. Any slope in the floor was too weak, rock fill mass covered weak material properties of the slope along so that circular 10 m mass sliding was avoided. However water drainage was almost varied. At that point, the ground-rock interface with a certain water holding structural feature needed the foam concrete application so that in the mass, reaching low shear strength planar levels were auger bored and tunneled and foam concrete -rock fill were applied as seen in Figure 5.

Investigation of Mass Movement and Slope Stability Analysis

In this study, the compression strength of the foam concrete applied landfills for the slopes in the area is illustrated in Figure 6. The stability analysis for S2 slope in terms of active and potential hazardous areas of landfill was carried out. The applied foam concrete on the soil surface improved the strength and eliminated hazard of sliding. According to studies of the stability without landfill, the land mass slope area there was relative movement designated as the active landslide area. Relative movements are determined by making use of tension cracks on the surface.

Belonging to the limit equilibrium factor of safety of slopes drawn in accordance with different sliding surfaces for the Bishop in GEO5, with the circular shift diagram Janbu and Fellenius methods showed over 1.3 safety factors. In the field the observed tension cracks showed the active movement. In the case of foam concrete landfill, the intrinsic parameters of the sliding surface will appear smaller than those safety values found with the parameters in the soils without the landfill. It is clear that safety value was between 1.5 and 1.8 with the landfill as seen from Figure 7. From this point of view, the foam concrete landfill of horizontal column improved water level under sliding surfaces and the safety factor values reached over 1.8 from instable value of 1.3. Deformation displacements at x axis showed hazardous levels of 30mm under the level of the ground water table and the sliding surface. However, over the sliding surface that displacement value remained very low 6 - 9 mm secured level.

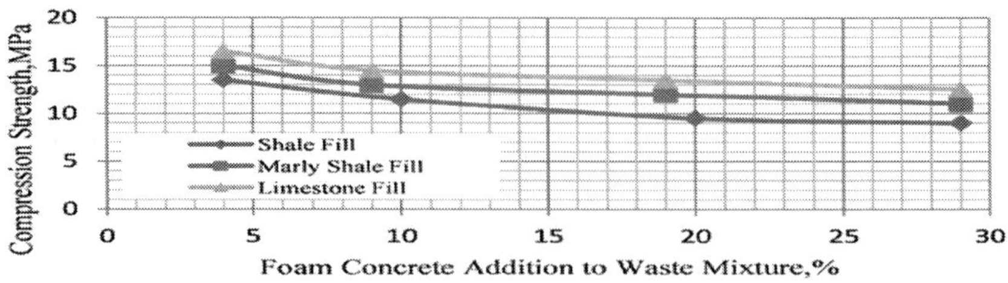

Figure 6: Compression strength of foam concrete landfill on hazardous landslide area.

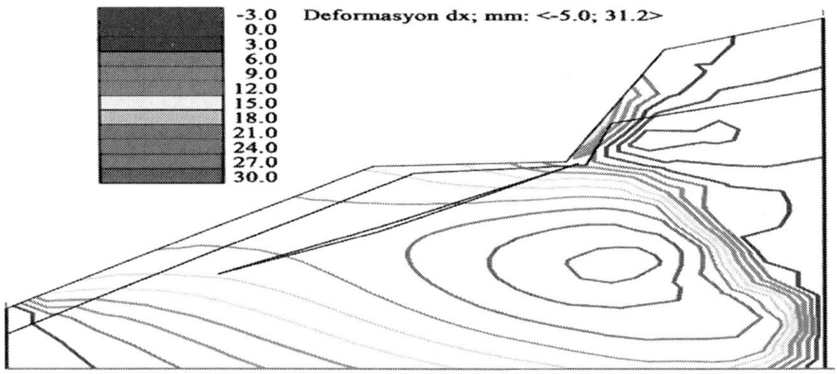

Figure 7: Slope stability, deformation values of foam concrete landfill on hazardous landslide area.

CONCLUSIONS

ırnak City located within the boundaries of four separate hill slopes and slope stability of soil samples should be taken from the site under construction. The geological and geotechnical properties investigated by field studies and laboratory experiments showed less plastic soil formations keeping water pore pressures.

Alluvium (Miocene), dissociated shale (Miocene) and the dissociated limestone (Miocene) are ranked propagating landslide hazardous manner of high slopes under construction sites.

The studied area of potential landslide hazard around the active site of stress, cracking displacement of relative motion could be observed, but changed regarding the field.

Soil samples performed on the laboratory test result in the slope material permeable that the cohesion value of 1.2 - 4.7 kPa, angle of internal friction of the 17.5° - 22.4° varied between unified soil classifications. Stability analysis performed in the light of this information, S1, S2 and S3 were unstable hillsides. By using foam concrete application, the hazardous slopes were concluded the stable condition. Various researches on foam concrete showed porous structure and advantageous points in concrete structures by use of fly ash.

As a result of the terrain views, especially in the months of May and June by the melting snow stream flow increase seen in the severe erosion of the slope formed on the heel and this effect was determined to have a negative impact on slope stability. Landfill material, or rock material or stream should be investigated and varied in water discharge effect.

In the experiments for different slope inclinations, especially in low-slope slope (1/1), very high security numbers (such as s2.0) were obtained. Therefore, by increasing the slope of the land slope stability calculations should be renewed. Slope safety number began to decrease with increasing safety and below 1.3 (approximately over 50°) slope inclinations.

Rocky slope stability analysis done for the ground, rocky land, which had been dissociated from the floor to act differently, next to the sliding movement of the rock fall incidents, such as being able to vary hazardous parameters. In the rocky terrain of slope stability analysis in order to solve problems with similar method of landfill may not be sufficient, however as a specially crafted construction works should be used.

High strength landfill mass improved the slope stability. However, low strength foam concrete landfill might result in higher water discharge and drier soil condition.

Plants, rain water infiltration into the mass simplifies and reduces water runoff and cause land sliding. This is an obstacle to the erosion of the mass. Reaching deep into the roots of the plants roots improves the mechanical balance of the masses.

Landfill porosity thus reduces the retaining force which stabilizes the slopes. For this reason, the vegetation of landslide hazardous areas is preventive enrichment an important parameter in the region. However, up to 30 m depth to the sliding surfaces of vegetation stability effect will be minimal. Weathering of rocks varies greatly into undergo, to the weakening of the bond between grains and leads to total extinction. In the study area weakened by weathering rocks are easily eroded and slope angle of inclination of the slope is changing with height. Dissociation observed in rocks in the study area also offers a negative contribution to stability problems.

For those reasons, geotechnical engineer can use virtual any classical slope stability programs, rock slope stability calculations in order to do construction safety.

As a result of this work performing of the geotechnical analysis, hazardous conditions would not be expected of a very large landslide. However, the urban living areas and the urban development areas at certain land slopes for the possibility of landslide hazard should need further technological improvements.

ACKNOWLEDGEMENTS

The author would like to thank Mining Engineering Department of Süleyman Demirel University.

REFERENCES

1. Bishop, A.W. (1955) The Use of the Slip Circle in the Stability Analysis of Earth Slopes. Geotechnique, 5, 7-17. http://dx.doi.org/10.1680/geot.1955.5.1.7
2. Cernica, J.N. (1995) Geotechnical Engineering: Soil Mechanics. John Wiley and Sons Inc., Canada.
3. Das, B.M. (1994) Principles of Geotechnical Engineering. PWS Publishing Company, USA.
4. Dramis, F. and Sorriso-Valvo, M. (1994) Deep-Seated Gravitational Slope Deformations, Related Landslides and Tectonics. Engineering Geology, 38, 231-243.http://dx.doi.org/10.1016/0013-7952(94)90040-X
5. Höek, E. (1970) Estimating the Stability of Excavated Slopes in Opencast Mines. Institution of Mining and Metallurgy, A105, A132.

6. Höek, E. and Bray, J.W. (1977) Rock Slope Engineering. Stephen Austin and Sons Ltd., Hertford.
7. Hoek, E. (2013) Practical Rock Engineering. Hoek Notes by Evert Hoek.http://www.rocscience.com
8. Hutchinson, J.N. (1995) Landslide Hazard Assessment. Keynote Paper. In: Bell, D.H., Ed., Landslides, Proceeding of 6th International Symposium on Landslides, Christchurch, Vol. 1, Balkema, Rotterdam, 1805-1841.
9. Lambe, W.T. and Whitman, R.V. (1969) Soil Mechanics. John Wiley and Sons, New York.
10. Langan, B.W., Weng, K. and Ward, M.A. (2002) Effect of Silica Fume and Fly Ash on Heat of Hydration of Portland Cement. Cement and Concrete Research, 32, 1045-1051.http://dx.doi.org/10.1016/S0008-8846(02)00742-1
11. Pa amehmetoğlu, A.G., Özgeno lu, A. and Karpuz, C. (1991) Rock Slope Stability. 2nd Edition, T.M.M.O.B Mining Engineering Bureau Publisher, Ankara.
12. Park, C.K., Noh, M.H. and Park, T.H. (2005) Rheological Properties of Cementitious Materials Containing Mineral Admixtures. Cement and Concrete Research, 35, 842-849.http://dx.doi.org/10.1016/j.cemconres.2004.11.002
13. Bouzoubaa, N., Bilodeau, A., Sivasundaram, V., Fournier, B. and Golden, D.M. (2004) Development of Ternary Blends of High-Performance Concrete. ACI Materials Journal, 101, 19-29.
14. Chindaprasirt, P., Homwuttiwong, S. and Sirivivatnanon, V. (2004) Influence of Fly Ash Fineness on Strength, Drying Shrinkage and Sulfate Resistance of Blended Cement Mortar. Cement and Concrete Research, 34, 1087-1092.http://dx.doi.org/10.1016/j.cemconres.2003.11.021
15. Hussin, M.W. and Awal, A.S.M.A. (1996) Palm Oil Fuel Ash—A Potential Pozzolanic Material in Concrete Construction. Proceedings of the International Conference on Urban Engineering in Asian Cities in the 21st Century, Bangkok, 20-23 November 1996, D361-D366.
16. Isaia, G.C., Gastaldini, A.L.G. and Moraes, R. (2003) Physical and Pozzolanic Action of Mineral Additions on the Mechanical Strength of High-Performance Concrete. Cement and Concrete

Composites, 25, 69-76. http://dx.doi.org/10.1016/S0958-9465(01)00057-9

17. Kiattikomol, K., Jaturapitakkul, C., Songpiriyakij, S. and Chutubtim, S. (2001) Study of Ground Coarse Fly Ashes with Different Finenesses from Various Sources as Pozzolanic Materials. Cement and Concrete Composites, 23, 335- 343. http://dx.doi.org/10.1016/S0958-9465(01)00016-6
18. Kim, H.K., Jeon, J.H. and Lee, H.K. (2012) Workability, and Mechanical, Acoustic and Thermal Properties of Lightweight Aggregate Concrete with a High Volume of Entrained Air. Construction and Building Materials, 29, 193- 200.http://dx.doi.org/10.1016/j.conbuildmat.2011.08.067
19. Chen, X., Yan, Y., Liu, Y.Z. and Hu, Z.H. (2014) Utilization of Circulating Fluidized Bed Fly Ash for Preparation of Foam Concrete. Construction and Building Materials, 54, 137-146. http://dx.doi.org/10.1016/j.conbuildmat.2013.12.020
20. Ranjani, G.I.S. and Ramamuthy, K. (2012) Behaviour of Foam Concrete under Sulphate Environments. Cement & Concrete Composites, 34, 825-834.http://dx.doi.org/10.1016/j.cemconcomp.2012.03.007
21. Liu, M.Y.J., Alengram, U.J., Jumaat, M.Z. and Mo, K.H. (2014) Evaluation of Thermal Conductivity, Mechanical and Transport Properties of Lightweight Aggregate Foamed Geopolymer Concrete. Energy and Buildings, 72, 238-245.
22. Xia, Y.Q., Yan, Y. and Hu, Z.H. (2013) Utilization of Circulating Fluidized Bed Fly Ash in Preparing Non-Autoclaved Aerated Concrete Production. Construction and Building Materials, 47, 1461-1467.
23. Sata, V., Jaturapitakkul, C. and Kiattikomol, K. (2007) From Pozzolan of Influence on Mechanical Properties of Various By-Product Materials High-Strength Concrete. Construction and Building Materials, 1589-1598.
24. Moulien, M. and Khelafib, H. (2008) Performance Characteristics of the Lightweight Aggregate Concrete Containing Natural Pozzolan. Building and Environment, 43, 31-36.
25. Demirbog, R., Orung, I. and Rose, R. (2001) Effects of Expanded Perlite Aggregate and Mineral Admixtures on the Compressive

Strength of Low-Density Concretes. Cement and Concrete Research, 31, 1627-1632.

26. Chen, B. and Liu, J.Y. (2008) Application of Experimental in Lightweight Concrete Admixtures with Mineral High Strength and Workability. Construction and Building Materials, 22, 655-659.

27. Anderson, M.G. and Richards, K.S. (1982) Slope Stability. John Wiley and Sons Ltd., New York.

28. Anonymous (2013) GEO5—Engineering Manuals—Part 1, Part 2.http://www.finesoftware.eu/geotechnical-software/

29. Anonymous (2009) GEO5—FEM—Theoretical Guide.http://www.finesoftware.eu/geotechnical-software/

30. Anonymous (2012) Şırnak City Bureau Reports. Şırnak.

31. ASTM (1990) Standard Test Method for Direct Shear Test of Soils under Consolidated Drained Condition. West Conshohocken, D3080-90.

32. ASTM (1985) Standart Specifications for Fly Ash and Raw or Calcined Natural Puzzolan for Use as Mineral Admixture in Portland Cement Concrete. ASTM, Philadelphia, ASTM C 618-85.

33. ASTM (1999) Standard Test Method for Time of Setting of Concrete Mixtures by Penetration Resistance. Pennsylvania, ASTM C 403.

34. Görög, P. and Török, Á. (2006) Stability Problems of Abandoned Clay Pits in Budapest. IAEG2006 P295, The Geological Society of London, London.

35. Görög, P. and Török, Á. (2007) Slope Stability Assessment of Weathered Clay by Using Field Data and Computer Modeling: A Case Study from Budapest. Natural Hazards and Earth System Sciences, 7, 417-422. www.nat-hazards-earth-syst-sci.net

36. Pruška, J. (2009) Comparison of Geotechnic Softwares—Geo FEM, Plaxis, Z-Soil. In: Vanícek, et al., Eds., XIII ECSMGE, Vol. 2, CGtS, Prague.

37. Sata, V., Jaturapitakkul, C. and Kiattikomol, K. (2007) Influence of Pozzolan from Various By-Product Materials on Mechanical Properties of High-Strength Concrete. Construction and Building Materials, 21, 1589-1598.

38. Vaneckova, V., Laurin, J. and Pruska, J. (2011) Sheeting Wall Analysis by the Method of Dependent Pressures. Geotech, Hanoi, 7.

39. Wiley, L. (1987) Slope Stability Geotechnical Engineering and Geomorphology. England.

Chapter 9

Improvement of Concrete Shear Wall Structures by Smart Materials

Mehdi Ghassemieh[1], Mohammad Reza Bahaari[1], Seyed Mohyeddin Ghodratian[1], and Seyed Ali Nojoumi[2]

[1]School of Civil Engineering, University of Tehran, Tehran, Iran

[2]Civil & Environmental Engineering Department, University of California, Los Angeles, USA

ABSTRACT

Smart materials have found numerous applications in many areas in civil engineering recently. One class of these materials is shape memory alloy (SMA) which exhibits several unique characteristics such as superelasticity and shape memory effect. Due to these characteristics, research efforts have been extended to use SMA in controlling civil structures. This paper investigates the effectiveness of SMA reinforcements in enhancing the behavior of shear walls, especially when subjected to seismic excitations. Two ordinary and coupled shear walls were introduced as reference structures and were modeled by ABAQUS software. For improving the seismic response of the shear walls, vertical SMA reinforcing bars were proposed to be implemented like conventional steel reinforcements, throughout the height of the structures and in every connecting beam in the coupled shear wall system. The one dimensional superelastic model of SMA material was implemented in the computer software using FORTRAN code. The dynamic response of the shear walls subjected to seismic loading was investigated through time history analyses under El-centro and Koyna records. The results showed that using superelastic SMA material instead of steel bars caused remarkable reduction in residual displacement for both ordinary and coupled shear walls. In addition, SMA reinforcements could significantly decrease the maximum deflection of the coupled shear wall system.

INTRODUCTION

Many multi-storey buildings contain shear walls around the elevator shafts and stairwells as lateral resisting systems. For the concrete shear wall systems, it is difficult to satisfy the very ductile behavior conditions. Therefore, such structures have often suffered damages caused by earthquake events. Shearing damage, bending damage, sliding and overturning damage are usually four kind of damage occur in concrete shear wall during earthquake. If concrete shear wall can attain their initial shape after an earthquake, then problems associated with permanent damage can be addressed. Sometimes, there are several openings in these shear walls and if two such openings are on opposite sides, deep coupling beams are supposed to interconnect the walls. These coupling beams are generally used as a means of dissipating energy during earthquakes through experiencing inelastic yielding. Due to their small span to depth ratio, they require highly congested reinforcement in order to achieve ductile behavior. Although dissipating energy through plastic hinging is a common practice in the design of multi-storey buildings, this practice usually results in significant residual displacements and the need to repair the structural elements after the earthquake. To address the shortcomings of current practices, a new design approach using smart materials such as shape memory alloy (SMA) is proposed.

SMA is one example of smart materials that exhibit several unique characteristics such as shape memory effect, superelasticity, and energy dissipation features. Due to these characteristics, Shape Memory Alloys have widely attracted attentions in passive control of structures in recent years. Dolce, et al. in a series of publications studied the effectiveness of SMA materials for use in seismic applications [1]. They also studied the implementation of various states of SMA materials for the use of special dampers in structures. They proposed different recentering or dissipating devices based on experimental results. Wilde et al. performed an analytical study of SMA-based seismic isolation system consist of laminated rubber bearing and superelastic SMA bars [2]. They conducted time history analysis with different excitation to compare the SMA-based bearing with conventional bearing with lead core. Dolce and Cardone experimentally investigated the proper choice of alloy, the effect of temperature, SMA size and loading rate and number of cycles [3]. Bruno and Valente showed the effectiveness of the use of SMA materials by analytical measures using simple pseudoelastic constitutive model for SMAs using damage index approach [4]. Baratta and Corbi analyzed the dynamics of a structural elastic-plastic frame, endowed with pseudoelastic SMA tendons [5]. DesRoches and Delemont evaluated the efficiency of using SMA restrainers to reduce the response of decks in a multi span simply supported bridge [6]. Masuda and Noori investigated the optimization of hysteretic characteristics of damping devices based on pseudoelastic SMAs [7]. DesRoches et al. experimentally evaluated the properties of superelastic Ni-Ti shape memory alloys under cyclic loading to assess their potential for applications in seismic resistant design and retrofit [8]. Abolmaali et al. compared the energy dissipative characteristics of bolted t stub connections using steel and shape memory alloy (SMA) fasteners [9]. Choi et al. proposed a new SMArubber bearing which is composed of a conventional elastomeric bearing and SMA wires wrapping the bearing in longitudinal direction [10]. A multilinear constitutive model developed by Motahari and Ghassemieh was adopted to capture the most common behaviors of SMA [11]. Czaderski et al. tested a reinforced concrete (RC) beam equipped with SMA material and compared it with conventional RC beam [12]. The results proved that by using shape memory alloys it is possible to produce a RC beam which has variable stiffness and strength. Saiidi and Wang presented the application of SMA bars instead of steel bars in plastic hinge zone of reinforced concrete bridge piers [13]. Motahari et al. also introduced a special SMA damper to have both re-centering and energy dissipating characteristics simultaneously [14]. Li et al. experimentally studied the behavior of smart concrete beams with embedded shape memory alloy bundles [15]. They used SMA bundles as actuators to achieve recovery force. Andrawes and DesRoches compared the efficiency of SMA restrainers with three other retrofit devices including conventional steel restrainers, metallic dampers and viscoelastic dampers [16]. Johnson et al. conducted a large scale testing program to evaluate the effect of SMA restrainer cable on the seismic performance of in-span hinges of multiple-frame concrete box girder bridge subjected to strong ground motion [17]. Rahman et al. investigated the effect of cross section geometry on the bending of a beam and also buckling of a column made of SMA through a numerical study [18]. Sharabash and Andrawes studied the application of SMAs as seismic passive damper devices for vibration mitigation of cable stayed bridges [19]. The feasibility of superelasticity in increasing ductility capacity and decreasing residual displacement of concrete bridge column was investigated by Saiidi et al. [20]. Ozbulut and Hurlebaus explored the effectiveness of SMA-rubber based isolation systems for seismic protection of bridges against near-field earthquakes. They also compare the performance of SMArubber based isolation systems with SMA-based sliding isolation system [21]. Kari et al. evaluated the effectiveness of a new dual bracing system for improving the seismic behavior of steel structures [22].

In this study the behavior of concrete shear walls reinforced with SMA bars is investigated. Finite element program, ABAQUS, was used in order to assess the response of the structures subjected to seismic loading. Two ordinary and coupled shear walls were introduced as reference structures and their seismic behavior with and without SMA reinforcement was evaluated through time history analyses.

SHAPE MEMORY ALLOY

Shape Memory Alloys are new class of metallic alloys that display multiple incomparable characteristics, based on martensitic phase transformation. SMAs are able to undergo large strains (8% - 10%) without leaving permanent deformations in the material. They can recover their initial shape at the end of the deformation process, instinctively (called superelasticity) or by heating (called shape memory effect) as shown in Figure 1. The most favorable characteristic of SMAs which is used in passive control of structures is superelastic behavior in which the material can recover large deformations of order of 8 percent while producing flag-shaped hysteresis. The second feature of the SMA is the shape memory effect. When the material is in Martensite form, application of stress leads to twinning of Martensite. By removal of stress the detwinning process begins and at the zero stress, some residual strain will be remained that can be recovered by heating the material above a specific temperature. Other desirable characteristics of SMAs are high energy dissipation capacity, stability of hysteresis loop and high fatigue resistance. Added to all those characteristics, the relatively high stiffness and strength makes SMA a promising material for control of structures in severe earthquakes. Although several alloys have the shape memory feature, the most widely used SMA in civil engineering applications is Nitinol, which is a combination of Nickel and Titanium. However, since Nitinol is a very expensive material compared to steel, it may not be economical to use it unless high energy dissipation is demanded.

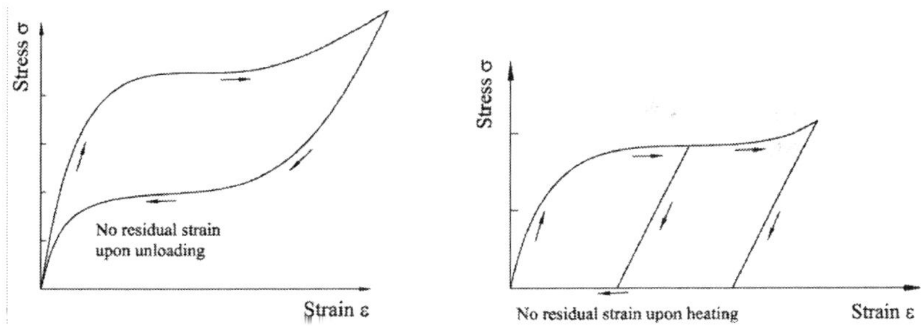

Figure 1: Stress-strain curves for SMAs: (a) superelasticity effect; (b) shape memory effect.

ANALYTICAL MODELS OF SHEAR WALLS

This study investigates the seismic performance of two concrete shear wall structures equipped with superelastic SMA. The first structure had the width of 5.0 m, height of 15.0 m and thickness of 0.3 m. The second structure was a coupled shear wall with two interconnected concrete walls. Each wall had the width of 3.0 m, height of 15.0 m and thickness of 0.3 m and each coupling beam was 0.6 m deep, 2.4 m length with 0.3 m width. Height between each level was fixed at 3.0 m. In order to attain the numerical behavior of the structures, finite element computer program (ABAQUS) was used in this study [23]. Finite Element models of the first and the second shear walls are presented in Figures 2 and 3, respectively. Concrete material had compressive strength of 32 MPa, Young's modulus of 30 GPa, poisson ratio of 0.2, and density of 25 kN/m^3. Concrete damage plasticity, developed by Lee and Fenves [24] was utilized for a proper material modeling of the concrete behavior in the numerical analyses. Lumped masses were placed at each node at story levels to represent the lateral inertial loads induced from the floor to the walls in time of earthquake excitations.

In order to grant better understanding of the dynamic behavior of the shear walls, a modal analysis was conducted for the reference structures by the ABAQUS computer program. Mode shapes and natural period of each mode are presented in Figures 4 and 5. As illustrated, the first three fundamental periods of the first shear wall were 0.726, 0.157 and 0.134 sec, respectively and the coupled shear wall had the first three periods of 0.828, 0.202 and 0.131 sec, respectively.

PROPOSED ENHANCEMENT TECHNIQUE

The proposed method uses SMAs as reinforcement in the concrete shear wall for the purposes of eliminating the residual displacement. For the upgraded structure, vertical SMA reinforcing rebars were proposed to be implemented like conventional steel reinforcements, throughout the height of the wall structures and in the all five connecting beams in coupled shear wall. No diagonal rebars were used as reinforcements in this research. For reaching the different percentages of reinforcements, the SMA and steel rebars were studied to posses various diameters and spacing. Since in the ABAQUS finite element program, the SMA mechanical behavior does not appear by default, thus the SMA material was implemented in the computer program by writing a subroutine using FORTRAN and attaching it to the main program as a subroutine material module.

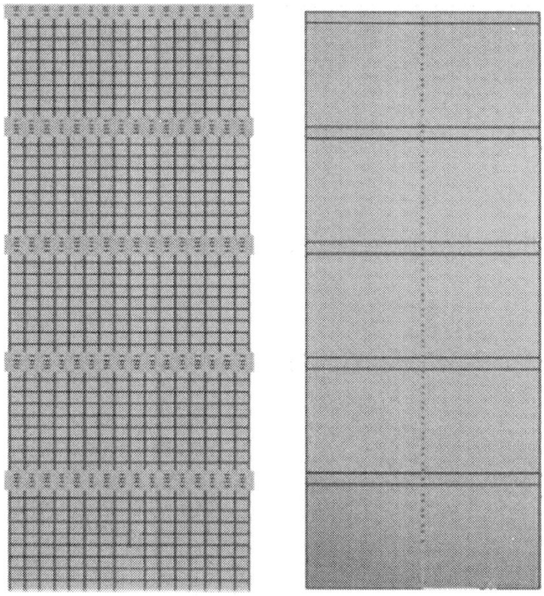

Figure 2: Finite element model of the ordinary shear wall without openings.

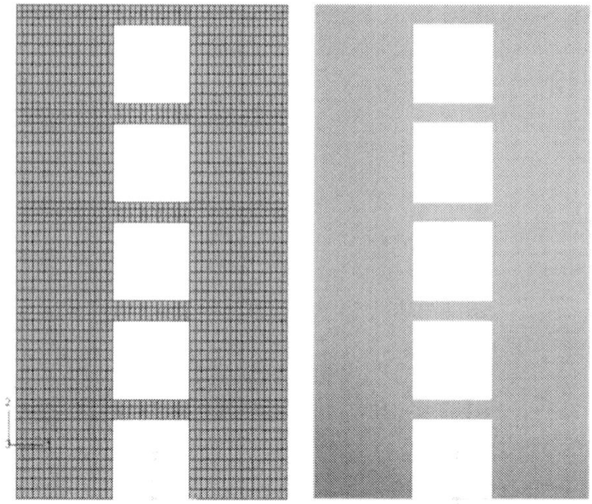

Figure 3: Finite element model of the coupled shear wall.

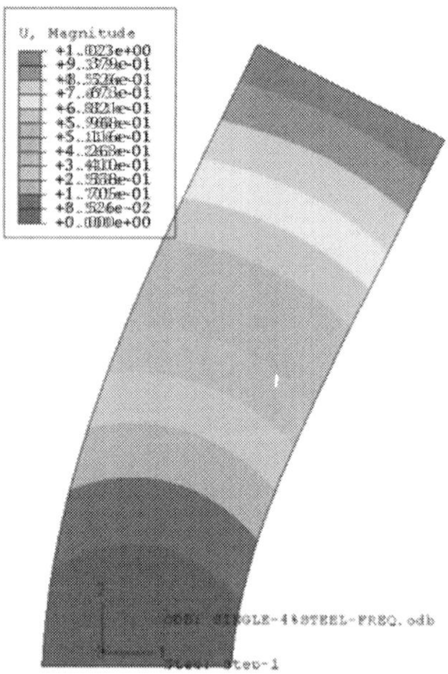

(a) 1st mode shape ($Tn_1 = 0.726$)

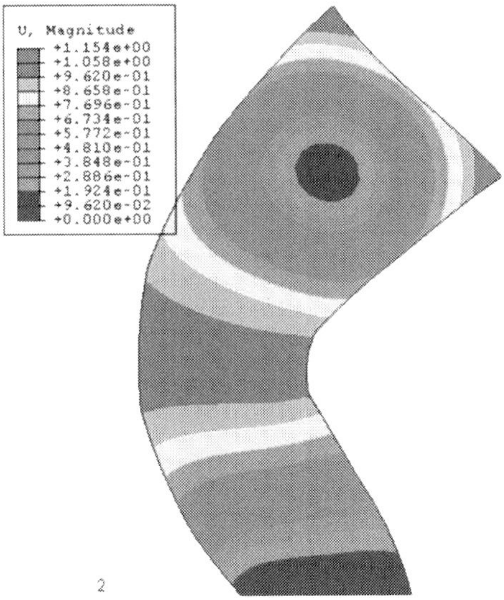

(b) 2nd mode shape ($Tn_2 = 0.157$)

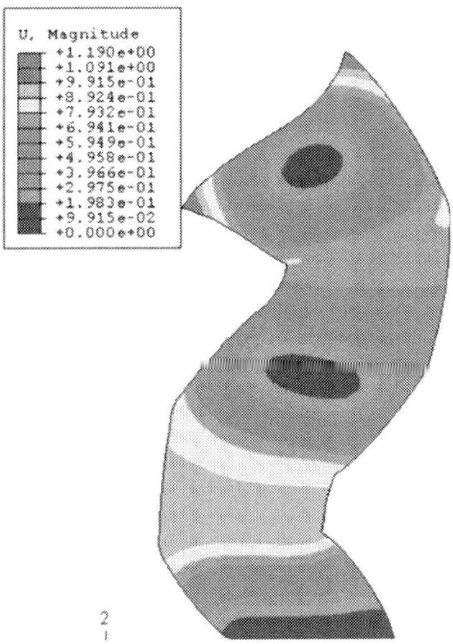

(c) 3rd mode shape ($Tn_3 = 0.073$)

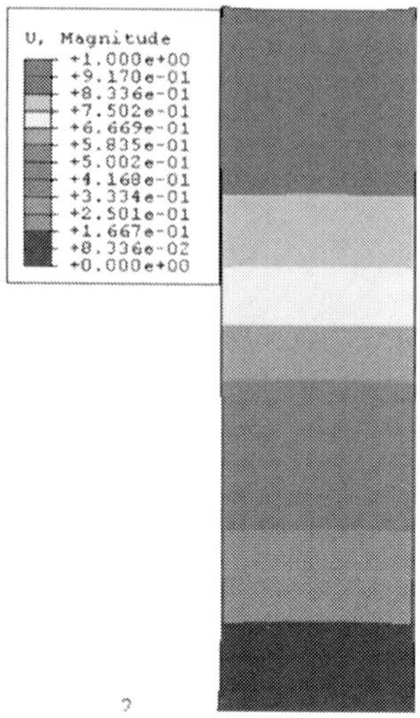

(d) 4th mode shape ($Tn_4 = 0.134$)

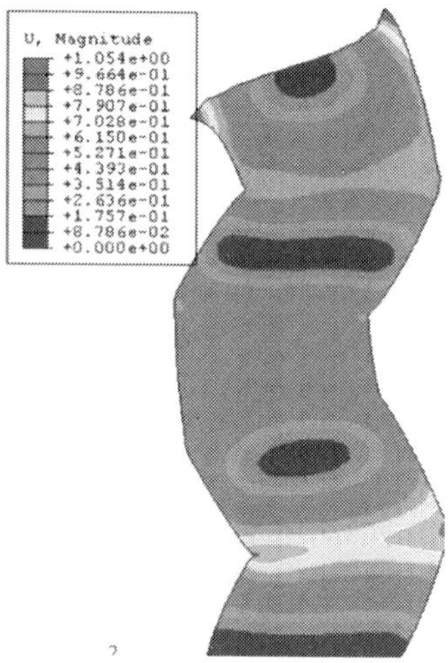

(e) 5th mode shape ($Tn_5 = 0.050$)

Figure 4: Mode shapes of the first shear wall.

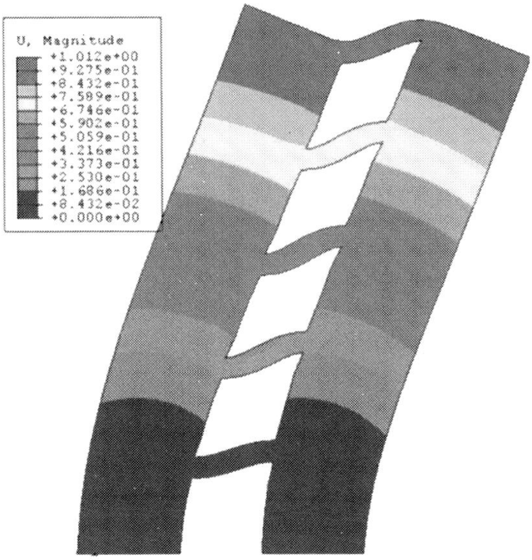

(a) 1st mode shape ($Tn_1 = 0.828$)

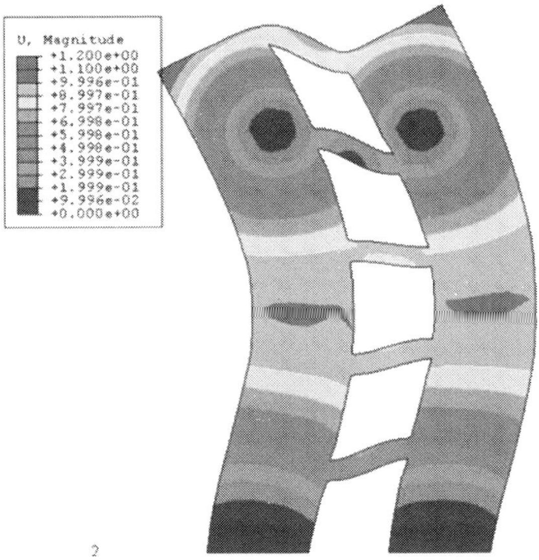

(b) 2nd mode shape ($Tn_2 = 0.202$)

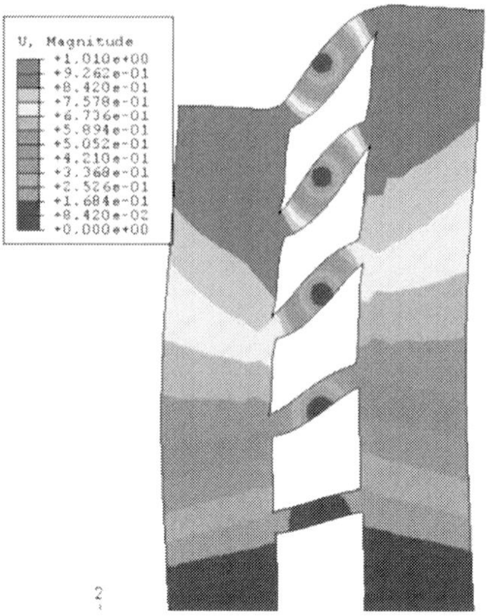

(c) 3rd mode shape ($Tn_3 = 0.131$)

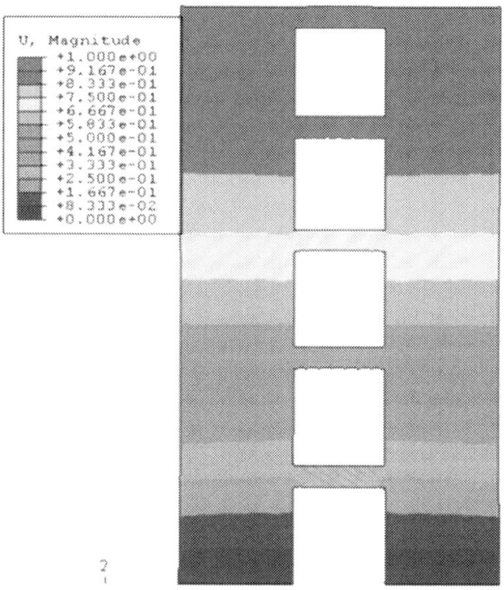

(d) 4th mode shape ($Tn_4 = 0.144$)

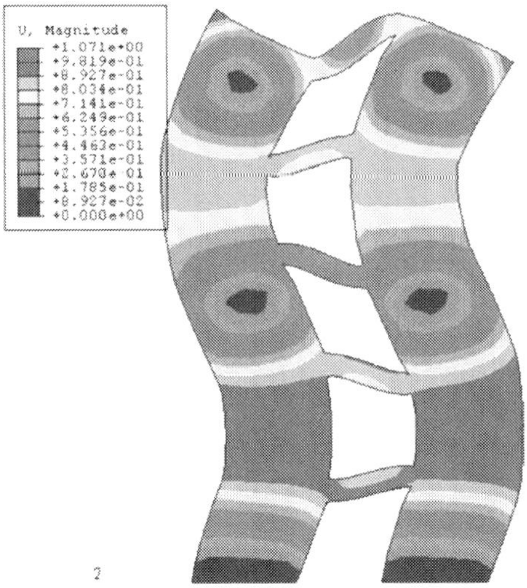

(e) 5th mode shape ($Tn_5 = 0.091$)

Figure 5: Mode shapes of the second coupled shear wall.

Since most civil engineering applications of SMA are related to the use of bars and wires, one dimensional phenomenological models are often considered suitable. Several researchers have proposed uniaxial models for SMA. Figure 6 shows the 1D-superelastic model of SMA material [11] implemented in the computer model. This model is capable of describing the constitutive behavior of superelastic SMAs at a constant temperature. The model requires 6 material parameters. The parameters used to define the material model are austenite to martensite starting stress ($_{am-s}$), austenite to martensite finishing stress ($_{am-f}$), martensite to austenite starting stress ($_{ma-s}$), martensite to austenite finishing stress ($_{ma-f}$), superelastic plateau strain length ($_{L}$), and modulus of elasticity (E_A). The SMA model represents an idealized behavior for SMA material where no strength degradation occurs during cycling and the residual deformation is taken zero at the end of each cycle. Further assumption is that austenite and martensite branches have the same modulus of elasticity (= 1). Previous studies have shown that such simplifications generally have negligible effect on the response [25]. Table 1 shows SMA mechanical properties defined in ABAQUS computer program as "User Implemented Material".

ANALYSES AND RESULTS

In this section, analyses were undertaken on two shear wall structures to evaluate the effectiveness of the SMA rebars in seismic performance of the concrete walls. The behavior of the structures subjected to earthquake excitations was investigated through dynamic time history analyses on the numerical models, and the response of the structure was attained by subjecting them to the El-centro and Koyna earthquake records. To show the significance of the SMA reinforcements in improving the seismic response, the story displacements at each floor level were obtained from time history analyses. The result of the new proposed system subjected to seismic excitations was then compared with those of the original concrete structure (without SMAs) and also assessed against models with different percentages of SMA reinforcements.

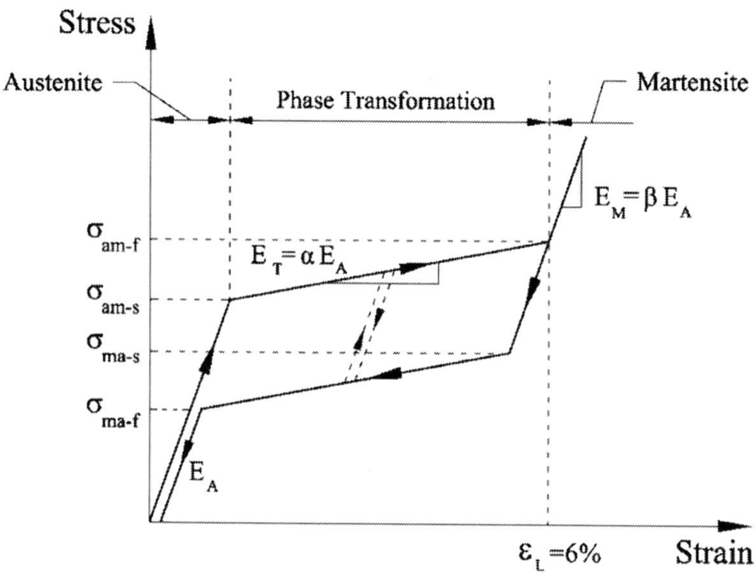

Figure 6: Stress-strain relationship of the super elastic SMA material.

Table 1: SMA mechanical properties

SMA Material Properties	Value
Modulus of elasticity(GPa)	40
Austenite to Martensite starting stress(MPa)	400
Austenite to Martensite finishing stress(MPa)	500
Martensite to Austenite starting stress(MPa)	300
Martensite to Austenite finishing stress(MPa)	200
Superelastic plateau strain length (%)	6

The First Shear Wall

Figures 7 and 8 compare the seismic response of the structure with steel or SMA reinforcement subjected to El-centro and Koyna records. The results showed that using superelastic SMA material instead of steel in concrete shear wall can significantly reduce the residual displacement. In particular, in the case of steel reinforcement, the concrete shear wall had residual displacements of 0.05 m and 0.08 m for El-centro and Koyna earthquakes, respectively. In the case of SMA reinforcement, the structure just experienced the residual displacements of 0.02 m and 0.02 m for El-centro and Koyna earthquakes, respectively. These results illustrated 60% and75% reduction in the residual displacement of the shear wall for El-centro and Koyna records, respectively. However, the results also showed that SMA reinforcements could not reduce the maximum deflection of ordinary shear walls meaningfully.

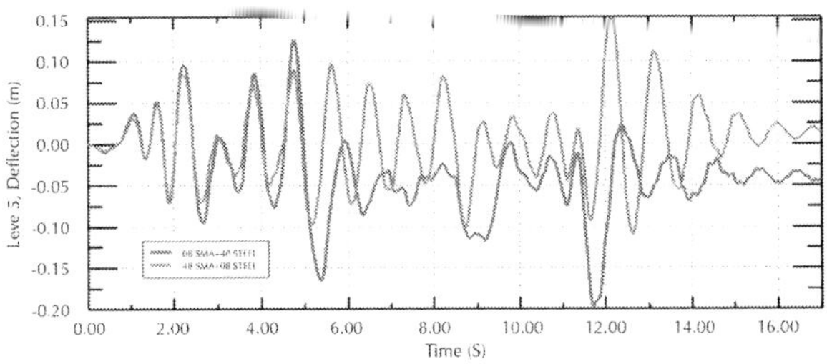

Figure 7: Tip deflection of the shear wall subjected to El-centro earthquake for different types of reinforcement.

The residual deformation of the shear wall subjected to El-centro earthquake for five stories is presented in Figure 9. As illustrated, the SMA reinforcement could successfully reduce the residual displacements compared to steel reinforcement for all five levels. This was mostly due to the superoelasticity feature of the SMA. In other words, in each cycle, shape memory alloys could restore most of the displacement and hence avoided the accumulation of residual displacement in the repeated cycles. Therefore, SMA limited the residual deformation at the end of the record. However, in the case of traditional steel reinforcement, yielding of material might accumulate plastic deformations in loading cycles and hence considerable residual deformation remained in the end of the earthquake.

Figure 10 displays the time histories of tip deflection for 4% and 6% SMA reinforcements in combination with 2% steel reinforcement. It can be observed that adding more SMA reinforcement to steel rebars could reduce both maximum and residual displacements. Specifically, by changing SMA reinforcement from 4% to 6%, the maximum tip displacement of the structure was decreased from 135 mm to 88 mm (46% reduction) and residual displacement was reduced from 30 mm to 15 mm (50% reduction).

The Second Coupled Shear Wall

Figure 11 compares the seismic response of the coupled shear wall with steel or SMA reinforcement under Koyna earthquake. The result showed that the coupled shear wall with SMA reinforcement experienced much lower level of deflection than the shear wall with steel reinforcement. In particular, in the case of steel reinforcement, the concrete wall had maximum deflection of 0.13 m, while in the case of SMA reinforcement, the structure just experienced the displacement of 0.07 m (i.e. 46% reduction in the maximum displacement). In addition, superelastic SMA reinforcement could significantly reduce the residual displacement of the concrete wall. Specifically, the residual displacement of the structure was reduced from 0.06 m in the original wall to 0.01 m in the controlled structure (i.e. 83% reduction in the residual displacement).

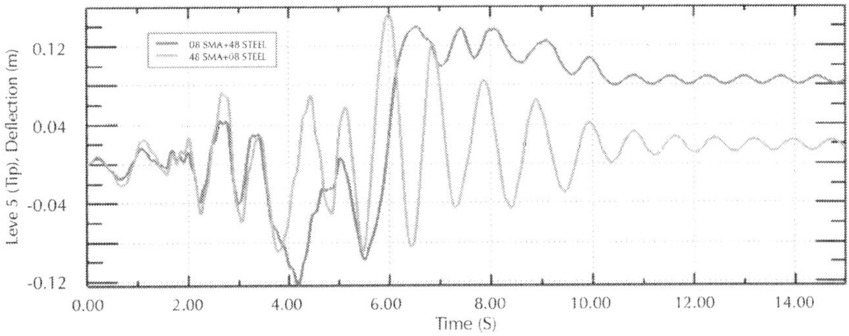

Figure 8: Tip deflection of the shear wall subjected to Koyna earthquake for different types of reinforcement.

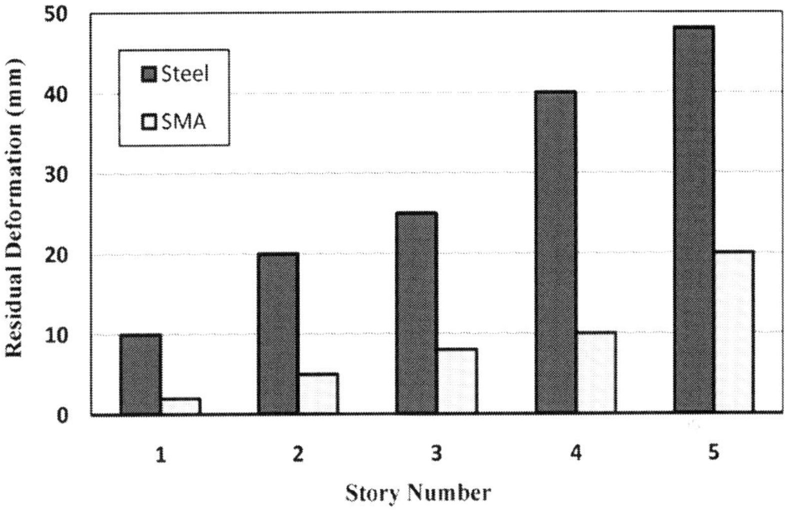

Figure 9: Residual deformation of the shear wall subjected to El-centro earthquake in five stories for steel or SMA reinforcements.

The maximum displacement of the coupled shear wall subjected to Koyna earthquake for five levels is illustrated in Figure 12. As it can be observed, using SMA reinforcement could meaningfully reduce the maximum displacement compared to steel reinforcement for all five stories especially for upper levels. This might be due to the fact that SMA bars were superelastic and had the ability to maintain their effective stiffness for repeated cycles while steel bars lost their effectiveness remarkably due to early yielding and caused large displacement in the response history.

Figure 13 displays the time histories of tip deflection for 0% steel and 4% SMA reinforcement and 1.5% steel and 2.5% SMA reinforcement. It can be observed that replacing 1.5% of shape memory alloy rebars by steel rebars did not affect the seismic response significantly and two concrete walls behave similarly under El-centro earthquake. However, since SMA is a very expensive material compared to steel, it may be economical to use it when the steel and SMA reinforcements are well proportioned.

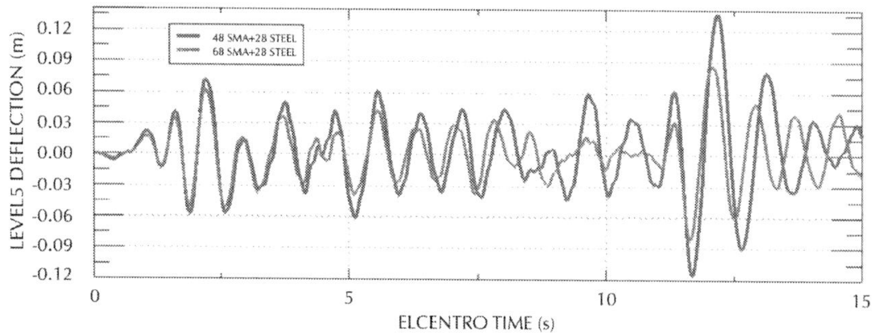

Figure 10: Tip deflection of the shear wall subjected to El-centro earthquake for different combination of SMA with steel reinforcement.

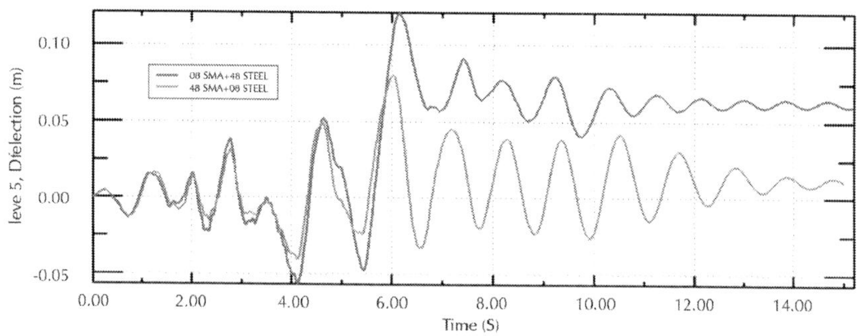

Figure 11: Tip deflection of the coupled shear wall subjected to Koyna earthquake for different types of reinforcement.

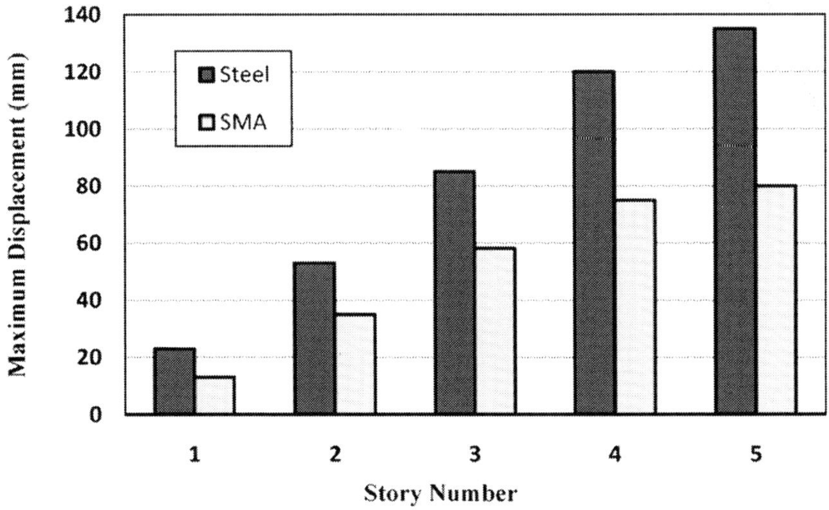

Figure 12: Maximum displacement of the coupled shear wall subjected to Koyna earthquake in five stories for steel or SMA reinforcements.

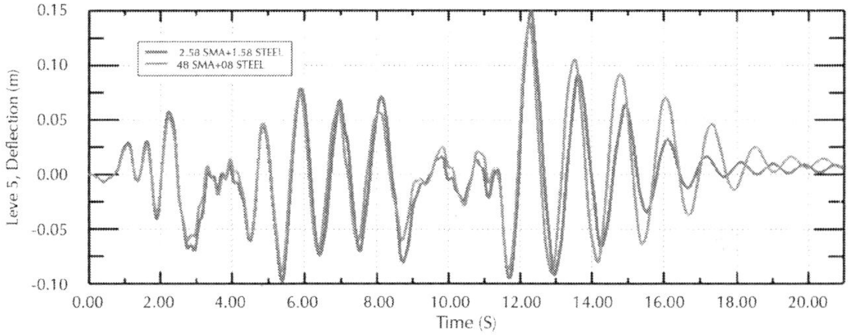

Figure 13: Tip deflection of the shear wall subjected to El-centro earthquake for different reinforcement combinations.

CONCLUSIONS

In this paper, an analytical study was conducted to evaluate the effectiveness of, one the latest smart material namely, shape memory alloy (SMA) in seismic improvement of concrete shear walls. Two types of shear walls such as ordinary type and coupled shear walls were introduced as reference structures and were modeled by ABAQUS computer program. Modal analyses were performed to capture the modal periods and mode shapes of the shear walls. For improving the seismic behavior of the shear walls, vertical SMA reinforcing bars were proposed to be implemented similar to conventional steel reinforcements, throughout the height of the structures and in all five connecting beams in coupled shear wall. The 1D-superelastic model of SMA material was implemented in the computer program by using FORTRAN as a subroutine material module. The dynamic response of the structures subjected to earthquake excitations was investigated through time history analyses subjected to El-centro and Koyna records.

In the case of the ordinary shear wall, as illustrated, using superelastic SMA material instead of steel rebars caused remarkable reduction in residual displacement. The results showed that SMA bars decreased the residual displacement by 60% and 75% for El-centro and Koyna records, respectively. However, the SMA reinforcements could not reduce the maximum deflection of the shear wall significantly. For the second coupled shear wall, it was observed that the structure with SMA reinforcement experienced lower level of maximum and residual displacements than the coupled shear wall with steel reinforcement. The results indicated that the SMA bars reduced the maximum and residual displacements up to 46% and 83%, respectively compared to the responses of the reference wall under Koyna earthquake.

In addition, different combinations of SMA reinforcement with steel reinforcement were modeled in both reference structures. The results showed that using such combinations of different reinforcement types improved the shear walls performance with an extent that depended on the percentage of SMA reinforcement. However, due to economical consideration, the steel and SMA reinforcements must be proportioned properly.

REFERENCES

1. M. Dolce, D. Cardone and R. Marnetto, "Implementation and Testing of Passive Control Devices Based on Shape Memory Alloys," Earthquake Engineering and Structural Dynamics,

Vol. 29, No. 7, 2000, pp. 945-968. doi:10.1002/1096-9845(200007)29:7<945::AID-EQE958>3.0.CO;2-#

2. K. Wilde, P. Gardoni and Y. Fujino, "Base Isolation System with Shape Memory Alloy Device for Elevated Highway Bridges," Engineering Structures, Vol. 22, 2000, pp. 222-229. doi:10.1016/S0141-0296(98)00097-2

3. M. Dolce and D. Cardone, "Mechanical Behavior of Shape Memory Alloys for Seismic Applications-2. Austenite NiTi Wires Subjected to Tension," International Journal of Mechanical Sciences, Vol. 43, No. 11, 2001, pp. 2657- 2677. doi:10.1016/S0020-7403(01)00050-9

4. S. Bruno and C. Valente, "Comparative Response Analysis of Conventional and Innovative Seismic Protection Strategies," Earthquake Engineering and Structural Dynamics, Vol. 31, No. 5, 2002, pp. 11067-11092. doi:10.1002/eqe.138

5. A. Baratta and O. Corbi, "On the Dynamic Behavior of Elastic-Plastic Structures Equipped with Pseudoelastic SMA Reinforcements," Computational Materials Science, Vol. 25, No. 2, 2002, pp. 1-13. doi:10.1016/S0927-0256(02)00245-8

6. R. DesRoches and M. Delemont, "Seismic Retrofit of Simply Supported Bridges Using Shape Memory Alloy," Engineering Structures, Vol. 24, No. 3, 2002, pp. 325- 332.doi:10.1016/S0141-0296(01)00098-0

7. A. Masuda and M. Noori, "Optimization of Hysteretic Characteristics of Damping Devices Based on Pseudoelastic Shape Memory Alloys," International Journal of Non-Linear Mechanics, Vol. 37, No. 8, 2002, pp. 1375- 1386. doi:10.1016/S0020-7462(02)00024-0

8. R. DesRoches, M. Delemont and J. McCormick, "Cyclical Properties of Superelastic Shape Memory Alloys," ASCE Journal of Structural Engineering, Vol. 130, No. 1, 2004, pp. 38-46. doi:10.1061/(ASCE)0733-9445(2004)130:1(38)

9. A. Abolmaali, J. Treadway, P. Aswath, F. K. Lu and E. McCarthy, "Hysteresis Behavior of T-Stub with Superelastic Shape Memory Fasteners," Journal of Constructional Steel Research, Vol. 62, No. 8, 2006, pp. 831-838. doi:10.1016/j.jcsr.2005.11.017

10. E. Choi, T.H. Nam, J.T. Oh and B.S. Cho, "An Isolation Bearing for Highway Bridges Using Shape Memory Alloys," Material

Science and Engineering A, Vol. 438-440, 2006, pp. 1081-1084. doi:10.1016/j.msea.2006.05.098

11. S. A. Motahari and M. Ghassemieh, "Multilinear Onedimensional Shape Memory Material Model for Use in Structural Engineering Applications," Engineering Structures, Vol. 29, No. 6, 2006, pp. 904-913. doi:10.1016/j.engstruct.2006.06.007
12. C. Czaderski, B. Hahnebach and M. Motavalli, "RC Beam with Variable Stiffness and Strength," Construction and Building Materials, Vol. 20, No. 9, 2006, pp. 824- 833.doi:10.1016/j.conbuildmat.2005.01.038
13. S. Saiidi and H. Wang, "Exploratory Study of Seismic Response of Concrete Columns with Shape Memory Alloys Reinforcement," ACI Structural Journal, Vol. 103, No. 3, 2006, pp. 436-443.
14. S. A. Motahari, M. Ghassemieh and S. A. Abolmaali, "Implementation of Shape memory Alloy Dampers for Passive Control of Structures Subjected to Seismic Excitations," Journal of Constructional Steel Research, Vol. 63, No. 12, 2007, 1570-1579.doi:10.1016/j.jcsr.2007.02.001
15. L. Li, Q. Li and F. Zhang "Behavior of Smart Concrete Beams with Embedded Shape Memory Alloy Bundles," Journal of Intelligent Material Systems and Structures, Vol. 18, No. 10, 2007, pp. 1003-1014. doi:10.1177/1045389X06071974
16. B. Andrawes and R. DesRoches, "Comparison between Shape Memory Alloy Seismic Restrainers and Other Bridge Retrofit Devices," Journal of Bridge Engineering, Vol. 12, No. 6, 2007, pp. 700-709. doi:10.1061/(ASCE)1084-0702(2007)12:6(700)
17. R. Johnson, J. E. Padgett, M. E. Maragakis, R. DesRoches and M. S. Saiidi, "Large Scale Testing of Nitinol Shape Memory Alloy Devices for Retrofitting of Bridges," Smart Material and Structures, Vol. 17, No. 3, 2008, pp. 1-28. doi:10.1088/0964-1726/17/3/035018
18. M. A. Rahman, S. R. Akanda and M. A. Hossain, "Effect of Cross Section Geometry on the Response of an SMA Column," Journal of Intelligent Material Systems and Structures, Vol. 19, No. 2, 2008, pp. 243-252.
19. A. M. Sharabash and B. Andrawes, "Application of Shape Memory Alloy Dampers in the Seismic Control of Cable-Stayed

Bridges," Engineering Structures, Vol. 31, No. 2, 2009, pp. 607-616. doi:10.1016/j.engstruct.2008.11.007

20. S. Saiidi, M. O'Brien and M. Sadrossadat-Zade, "Cyclic Response of Concrete Bridge Columns Using Superelastic Nitinol and Bendable Concrete," ACI Structural Journal, Vol. 106, No. 1, 2009, pp. 69-77.
21. O. E. Ozbulut and S. Hurlebaus, "Seismic Assessment of Bridge Structures Isolated by a Shape Memory Alloy/ Rubber-Based Isolation System," Smart Materials and Structures, Vol. 20, No. 1, 2011, pp. 1-15. doi:10.1088/0964-1726/20/1/015003
22. A. kari, M. Ghassemieh and A. Abolmaali, "A New Dual Bracing System for Improving the Seismic Behavior of Steel Structures," Smart Materials and Structures, Vol. 20, No. 12, 2011, pp. 1-35. doi:10.1088/0964-1726/20/12/125020
23. Abaqus, Inc., ABAQUS User Manual, V6.4, 2006.
24. J. Lee and G. L. Fenves, "A Plastic-Damage Model for Cyclic Loading of Concrete Structures," ASCE Journal of Engineering Mechanics, Vol. 124, No. 8, 1998, pp. 892- 900.doi:10.1061/(ASCE)0733-9399(1998)124:8(892)
25. B. Andrawes and R. DesRoches, "Sensitivity of Seismic Applications to Different Shape Memory Alloy Models," Journal of Engineering Mechanics, Vol. 134, No. 2, 2008, pp. 173-183. doi:10.1061/(ASCE)0733-9399(2008)134:2(173)

Chapter 10

Comparison of Design and Analysis of Concrete Gravity Dam

Md. Hazrat Ali, Md. Rabiul Alam, Md. Naimul Haque, and Muhammad Jahangir Alam

Department of Civil Engineering, Chittagong University of Engineering & Technology, Chittagong, Bangladesh

ABSTRACT

Gravity dams are solid concrete structures that maintain their stability against design loads from the geometric shape, mass and strength of the concrete. The purposes of dam construction may include navigation, flood damage reduction, hydroelectric power generation, fish and wildlife enhancement, water quality, water supply, and recreation. The design and evaluation of concrete gravity dam for earthquake loading must be based on appropriate criteria that reflect both the desired level of safety and the choice of the design and evaluation procedures. In Bangladesh, the entire country is divided into 3 seismic zones, depending upon the severity of the earthquake intensity. Thus, the main aim of this study is to design high concrete gravity dams based on the U.S.B.R. recommendations in seismic zone II of Bangladesh, for varying horizontal earthquake intensities from 0.10 g - 0.30 g with 0.05 g increment to take into account the uncertainty and severity of earthquake intensities and constant other design loads, and to analyze its stability and stress conditions using analytical 2D gravity method and finite element method. The results of the horizontal earthquake intensity perturbation suggest that the stabilizing moments are found to decrease significantly with the increment of horizontal earthquake intensity while dealing with the U.S.B.R. recommended initial dam section, indicating endanger to the dam stability, thus larger dam section is provided to increase the stabilizing moments and to make it safe against failure. The vertical, principal and shear stresses obtained using ANSYS 5.4 analyses are compared with those obtained using 2D gravity method and found less compares to 2D gravity method, except the principal stresses at the toe of the gravity dam for 0.10 g - 0.15 g. Although, it seems apparently that smaller dam section may be sufficient for stress analyses using ANSYS 5.4, it would not be possible to achieve the required factors of safety with smaller dam section. It is observed during stability analyses that the factor of safety against sliding is satisfied at last than other factors of safety, resulting huge dam section to make it safe against sliding. Thus, it can be concluded that it would not be feasible to construct a concrete gravity dam for horizontal earthquake intensity greater than 0.30 g without changing other loads and or dimension of the dam and keeping provision for drainage gallery to reduce the uplift pressure significantly.

INTRODUCTION

Basically, a gravity concrete dam is defined as a structure, which is designed in such a way that its own weight resists the external forces. It is primarily the weight of a gravity dam which prevents it from being overturned when subjected to the thrust of impounded water [1]. This type of structure is durable, and requires very little maintenance. Gravity dams typically consist of a nonoverflow section(s) and an overflow section or spillway. The two general concrete construction methods for concrete gravity dams are conventional placed mass concrete and RCC. Gravity dams, constructed in stone masonry, were built even in ancient times, most often in Egypt, Greece, and the Roman Empire [2,3].

However, concrete gravity dams are preferred these days and mostly constructed. They can be constructed with ease on any dam site, where there exists a natural foundation strong enough to bear the enormous weight of the dam. Such a dam is generally straight in plan, although sometimes, it may be slightly curve. The line of the upstream face of the dam or the line of the crown of the dam if the upstream face in sloping, is taken as the reference line for layout purposes, etc. and is known as the "Base line of the Dam" or the "Axis of the Dam". When suitable conditions are available, such dams can be constructed up to great heights. The ratio of base width to height of high gravity dams is generally less than 1:1.

A typical cross-section of a high concrete gravity dam is shown in Figure 1.

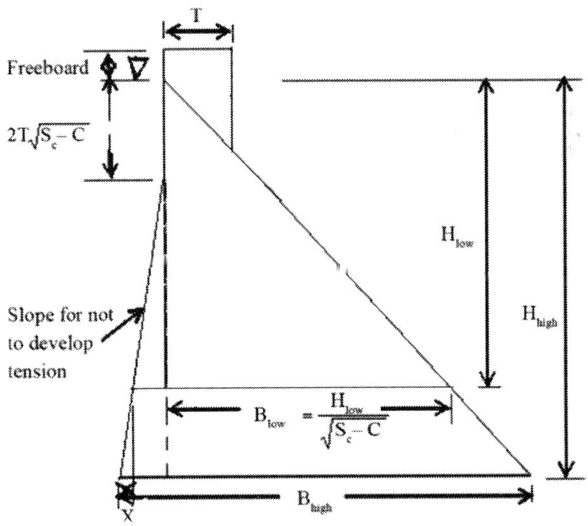

Figure 1: Typical section of a high concrete gravity dam.

The upstream face may be kept throughout vertical or partly slanting for some of its length. A drainage gallery is generally provided in order to relieve the uplift pressure exerted by the seeping water. Purposes applicable to dam construction may include navigation, flood damage reduction, hydroelectric power generation, fish and wildlife enhancement, water quality, water supply, and recreation.

Many concrete gravity dams have been in service for over 50 years, and over this period important advances in the methodologies for evaluation of natural phenomena hazards have caused the design-basis events for these dams to be revised upwards. Older existing dams may fail to meet revised safety criteria and structural rehabilitation to meet such criteria may be costly and difficult. The identified causes of failure, based on a study of over 1600 dams [4] are: Foundation problems (40%), Inadequate spillway (23%), Poor construction (12%), Uneven settlement (10%), High poor pressure (5%), Acts of war (3%), Embankment slips (2%), Defective materials (2%), Incorrect operation (2%), and Earthquakes (1%). Other surveys of dam failure have been cited by [5], who estimated failure rates from 2 ´ 10^{-4} to 7 ´ 10^{-4} per damyear based on these surveys.

LOADS

In the design of gravity concrete, it is essential to determine the loads required in the stability and stress analyses. The forces which may affect the design are: 1) Dead load or stabilizing force; 2) Headwater and tailwater pressures; 3) Uplift; 4) Temperature; 5) Earth and silt pressures; 6) Ice pressure; 7) Earthquake forces; 8) Wind pressure; 9) Subatmospheric pressure; 10) Wave pressure, and 11) Reaction of foundation.

The seismic safety of such dams has been a serious concern since damage to the Koyna Dam in India in 1967 which has been regarded as a watershed event in the development of seismic analysis and design of concrete gravity dams all over the world. It is essential that those responsible must implement policies and procedures to ensure seismic safety of dams through sound professional practices and state-of-the-art in related technical areas. Seismic safety of dams concerns public safety and therefore demands a higher degree of public confidence. The estimations and descriptions of various forces are provided briefly in the following sections.

Water Pressure

Water pressure (P) is the most major external force acting on gravity dams. The horizontal water pressure exerted by the weight of water stored on the upstream and downstream sides of the dam can be estimated from the rule of hydrostatic pressure distribution and can be expressed by

$$P = \frac{1}{2}\gamma_w H^2 \tag{1}$$

where, H is the depth of water and g_w is the unit weight of water.

Uplift Pressure

Water seepage through the pores, cracks and fissures of the foundation materials, and water seepage through dam body and then to the bottom through the joints between the body of the dam and its foundation at the base exert an uplift pressure on the base of the dam. According to the [6], the uplift pressure intensities at the heel and toe of the dam should be taken equal to their respective hydrostatic pressures and joined the intensity ordinates by a straight line. When drainage galleries are provided to relieve the uplift, the recommended uplift at the face of the gallery is equal to the hydrostatic pressure at toe plus $1/3^{rd}$ of the difference between the hydrostatic pressures at the heel and the toe, respectively.

Earthquake Forces

An earthquake produces waves, which are capable of shaking the earth upon which the gravity dams rest, in every possible direction. The effect of an earthquake is, therefore, equivalent to imparting acceleration to the foundations of the dams in the direction in which the wave is traveling at the moment.

Generally, an earthquake induces horizontal acceleration (a_h) and vertical acceleration (a_v). The values of these accelerations are generally expressed as percentage of the acceleration due to gravity (g), i.e., a = 0.10 g or 0.20 g, etc. On an average, a value of a equal to 0.10 to 0.15 g is generally sufficient for high dams in seismic zones. In extremely seismic regions and in conservative designs, even a value up to 0.30 g may sometimes be adopted [7].

Earthquake loadings should be checked for horizontal as well as vertical earthquake accelerations. While earthquake acceleration might take place in any direction, the analysis should be performed for the most unfavorable direction.

The earthquake loadings used in the design of concrete gravity dams are based on design earthquakes and sitespecific motions determined from seismological evaluation. At a minimum, a seismological evaluation should be performed on all projects located in seismic zones 1, 2, and 3 of Bangladesh [8], depending upon the severity of earthquakes.

The seismic coefficient method of analysis should be used in determining the resultant location and sliding stability of dams. In strong seismicity areas, a dynamic seismic analysis is required for the internal stress analysis.

Effect of Vertical Acceleration (($_v$)

A vertical acceleration may either act downward or upward. When it acts in the upward direction, then the foundation of the dam will be lifted upward and becomes closer to the body of the dam, and thus the effective weight of the dam will increase and hence, the stress developed will increase.

When the vertical acceleration acts downward, the foundation shall try to move downward away from the dam body; thus, reducing the effective weight and the stability of the dam, and hence is the worst case for design. The net effective weight of the dam is given by

$$W - \frac{W}{g} k_v g = W\left(1 - k_v\right) \tag{2}$$

where, W is the total weight of the dam, k_v is the fraction of gravity adopted for vertical acceleration, such as 0.10 or 0.20, etc.

In other words, vertical acceleration reduces the unit weight of the dam material and that of water to $(1 - k_v)$ times their original unit weights.

Effects of Horizontal Acceleration (($_h$)

The horizontal acceleration may cause 1) hydrodynamic pressure, and 2) horizontal inertia force.

- Hydrodynamic Pressure: Horizontal acceleration acting towards the reservoir causes a momentary increase in the water pressure, as the foundation and dam accelerate towards the reservoir and the water resists the movement owing to its inertia. According to [9], the amount of this hydrodynamic force (P_e) is given by

$$P_e = 0.726 C_m k_h \gamma_w H^2 \tag{3}$$

where, C_m = maximum value of pressure coefficient for a given constant slope =0.735(q°/90), where θ is the angle in degree, which the upstream face of the dam makes with the horizontal; k_h = fraction of gravity adopted for horizontal acceleration (a_h) such as.a_h=k_h g.

The moment of this force about the base is given by

$$M_e = 0.412 P_e H \tag{4}$$

- Horizontal Inertia Force: In addition to exerting the hydrodynamic pressure, the horizontal acceleration produces an inertia force into the body of the dam. This force is generated to keep the body and the foundation of the dam together as one piece. The direction of the produced force will be opposite to the acceleration imparted by the earthquake.

Since an earthquake may impart either upstream or downstream acceleration, it is needed to choose the direction of this force in the stability analysis of dam structure in such a way that it produces most unfavorable effects under the considered conditions. For example, when the reservoir is full, this force will produce worst results if it is additive to the hydrostatic water pressure, thus acting towards the downstream (i.e., when upstream earthquake acceleration towards the reservoir is produced). When the reservoir is empty, this force would produce worst results, if considered to be acting upstream (i.e., when earthquake acceleration moving towards the downstream is produced.

The horizontal inertia force is given by

$$\frac{W}{g}\alpha_h = \frac{W}{g}k_h g = Wk_h \tag{5}$$

where, k_h is the fraction of gravity adopted for horizontal acceleration, such as 0.10 or 0.20, etc.

This force should be considered to be acting at the center of gravity of the mass, regardless of the shape of the cross-section, and it acts horizontally downstream in worst cases, i.e., for reservoir full case.

Silt Pressure

Generally, silt gets deposited at the upstream face of the dam. If h is the height of silt deposited, then the force exerted by this silt in addition to external water pressure, can be represented by Rankine's formula as

$$P_{\text{silt}} = \frac{1}{2}\gamma_{\text{sub}} h^2 K_a \tag{6}$$

where, K_a is the coefficient of active earth pressure of silt =(1-sinφ)/(1+sinφ), where φ is the angle of internal friction of soil (cohesion is neglected); and g_{sub} is the submerged unit weight of silt material.

Wave Pressure

Waves are generated on the surface of the reservoir by the blowing winds, ships, etc., which cause a pressure towards the downstream side. Wave pressure depends upon the wave height (h_w) and the total force due to wave action (P_w) can be expressed as

$$P_w = 19.62h_w^2 \tag{7}$$

This force acts at a distance $\frac{3}{8}h_w$ below the reservoir surface.

Ice Pressure

The ice which may be formed on the water surface of the reservoir in cold countries may sometimes melt and expand. The dam face has then to resist the thrust exerted by the expanding ice. This force acts linearly along the length of the dam and at the reservoir level. Estimates of the thrust force vary according to the anticipated ice thickness, and range from 2.4 ´ 10^5 to 14.4 ´ 10^5 N/$m^2$of contact with the vertical face of the dam [10]. On an average, a value of 500 kN/m^2 may be allowed for ordinary conditions.

Weight of the Dam

The weight of the dam body and its foundation is the major resisting force. In two dimensional analysis of a gravity dam, the cross-section can be divided into rectangles and triangles. The resultant of all the downward forces will represent the total weight of the dam acting at the center of gravity of the dam.

PROFILE OF A DAM FROM PRACTICAL CONSIDERATIONS

The elementary profile of a gravity dam (i.e., triangle with maximum water surface at apex) is only a theoretical profile. Certain changes will have to be made in this profile in order to cater to the practical needs. These needs are: 1) providing a straight top width for a road construction over the top of the dam, and 2) providing a free-board above the top water surface so that water may not spill over the top of the dam due to wave action, etc.

Top Width

The top width (T) of the high dam should be sufficient to provide carriage-way and to withstand earthquake shocks and wave actions, and this may be selected like large earthen dam as

$$T = 1.65\left(H + 1.5\right)^{1/3} \tag{8}$$

The most economical top width without considering earthquake forces varies between 6 and 10 m and is generally taken approximately equal to $\sqrt{H}$.

Freeboard

Nowadays a freeboard equal to 4% - 5% of the dam height is provided. The section of gravity dam should be chosen in such a way that it is the most economic section and satisfies all the conditions and requirements of stability. Hence, after the section of the dam has been arrived at, the stability analysis for the dam must be carried out.

LOW AND HIGH CONCRETE GRAVITY DAMS

The height of the dam to be constructed should be checked to ensure whether it is a low gravity dam or a high gravity dam. If the height of the dam is less than that given by $f / (\gamma_w (S_c + 1))$, where f is the permissible compressive stress of the dam material; S_c is the specific gravity of the dam material; then the dam will be a low gravity dam, otherwise it will be a high gravity dam (Figure 1).

When the height of the dam exceeds the height of the low gravity dam (i.e., high dam), the upper height equal to low gravity dam can be designed as low gravity dam and its remaining height can be designed as follows [7]:

The total base width required for high gravity dam can be given by

$$B_{\text{high}} = \sqrt{\frac{\lambda_w H_{\text{high}}^3}{f}\left(1+\frac{\lambda_w^2 H_{\text{high}}^4}{4W}\right)} \tag{9}$$

and the increase in length at u/s face of high gravity dam can be determined by

$$\frac{\gamma_w S_c\left(B_{\text{high}} - B_{\text{low}}\right)}{24}\begin{bmatrix} 3B_{\text{low}}^2 - B_{\text{high}}^2 \\ +6\left(B_{\text{low}} + B_{\text{high}}\right) \\ +2B_{\text{low}}B_{\text{high}} \end{bmatrix} - \frac{\gamma_w x}{12}\left(H_{\text{low}} + H_{\text{high}}\right)\left(2B_{\text{high}} - 3x\right) - W_{\text{high}}\left(\frac{B_{\text{high}} - B_{\text{low}}}{3} - x\right) = 0 \tag{10}$$

where, W is the total weight of dam portion and water above the base of the high dam, B_{low} is the base width of low gravity dam, H_{high} is the base width of high gravity dam, x is the increment of length at u/s face, H_{low} is the height of maximum water level above the base of low gravity dam, H_{high} is the height of maximum water level above the base of high gravity dam, and W_{high} is the total weight of dam portion and water above the base of high gravity dam.

DESIGN CRITERIA

The design and evaluation of hydraulic structures for earthquake loading must be based on appropriate criteria that reflect both the desired level of safety and the choice of the design and evaluation procedures. The first requirement is to establish design earthquake ground motions to be used as the seismic input by giving due consideration to the consequences of failure and the designated operational function. Then the response of the structure to this seismic input must be calculated taking into account the significant interactions with the rock, soil, or pile foundation as well as with the impounded, or surrounding and contained water. The analysis should be formulated using a realistic idealization of the structure-water-foundation system, and the results are evaluated in view of the limitations, assumptions, and uncertainties associated with the seismic input and the method of analysis.

The design of a gravity dam should be checked for two cases, i.e., 1) Reservoir Full Case, and 2) Reservoir Empty Case. When the reservoir is full, the major forces acting are: weight of the dam, external water pressure, uplift pressure, and earthquake forces in serious seismic zones. The minor forces are: silt pressure, ice pressure in cold countries and wave pressure. For the most conservative design and from purely theoretical point of view, a situation may arise when all the forces may act together. For reservoir empty case, empty reservoir with a horizontal earthquake force produced towards the u/s has to be checked for non-development of tension at toe. In this study, water pressure due to maximum pool level, extreme uplift pressure without any reduction due to drainage, tailwater pressure up to a certain height, and earthquake forces, will be considered in the design of gravity dam.

MODES OF FAILURES AND STRUCTURAL STABILITY OF GRAVITY DAMS

A gravity dam may fail due to 1) overturning/rotation about the toe; 2) crushing; 3) development of tension, causing ultimate failure by crushing; and 4) shear failure called sliding.

Overturning/Rotation about the Toe

If the resultant of all the forces acting on a dam at any of its sections passes outside the toe, the dam shall rotate and overturn about the toe. The factor of safety against overturning (F.S.O.) generally varies from 1.5 to 2.

Compression or Crushing

A dam may fail by the failure of its materials, i.e., the compressive stresses produced may exceed the allowable stresses, and the dam material may get crushed. The vertical direct stress distribution at the base is given by

$$p_{\frac{max}{min}} = \frac{\sum V}{B}\left(1 \pm \frac{6e}{B}\right) \tag{11}$$

where, e is the eccentricity of the resultant force from the center of the base, the maximum value of which can be permitted on either side of the center of the base is equal to B/6; SV is the total vertical force; and B is the base width of the dam.

Tension

Masonry and concrete gravity dams are usually designed in such a way that no tension is developed anywhere, because these materials cannot withstand sustained tensile stresses. For achieving economy in designs of very high gravity dams, certain amount of tension may be permitted under severest loading condition. The maximum permissible tensile stress for high concrete gravity dams, under worst loadings, may be taken as 500 kN/m^2.

Sliding

Stability against sliding and shearing through a certain section through the dam across the foundation or along cracks in the foundation is of utmost importance. That is why it needs to be examined with special attention [11,12]. Sliding or shear failure will occur when the net horizontal force above any plane in the dam or at the base of the dam exceeds the frictional resistance developed at that level. The factor of safety against sliding (F.S.S.) is given by

$$\text{F.S.S.} = \frac{\mu \sum V}{\sum H} \tag{12}$$

where, mSV is the shear resistance in which SV is the total vertical forces; m is the coefficient of friction between the dam base and foundation, which varies from 0.65 - 0.75; and SH is the total external horizontal forces.

In low dams, the safety against sliding should be checked only for friction, but in high gravity dams, for economic precise designs, the shear strength of the joint, which is an additional shear resistance, must also be considered. If this shear resistance of the joint is considered, then the equation for factor of safety against sliding which is measured by shear friction factor (S.F.F.) becomes

$$\text{S.F.F.} = \frac{\mu \sum V + Bq}{\sum H} \tag{13}$$

where, q is the average shear strength of the joint which varies from about 1400 kN/m^2 for poor rocks to about 4000 kN/m^2 for good rocks.

Principal and Shear Stresses

The vertical stress intensity, p_{max} or p_{min} determined using Equation (11) is not the maximum direct stress produced anywhere in the dam. The maximum normal stress will, indeed, be the major principal stress that will be generated on the major principal plane. The principal (s) and shear (t) stresses at the toe and heel of gravity dam can be expressed [7] by

$$\sigma_{\text{toe}} = p_v \sec^2 \alpha - \left(p' - p_e'\right) \tan^2 \delta \tag{14}$$

$$\sigma_{\text{heel}} = p_v \sec^2 \varphi - \left(p + p_e\right) \tan^2 \theta \tag{15}$$

$$\tau_{(toe)} = \left[p_v - \left(p' - p_e' \right) \right] \tan \delta \quad (16)$$

$$\tau_{(heel)} = \left[p_v - \left(p + p_e \right) \right] \tan \theta \quad (17)$$

where, δ is the angle which the downstream face of the dam makes with the vertical, θ is the angle which the upstream face makes with the vertical, p_v is the intensity of uplift pressure, p is the minor principal stress at the heel, p_e is the hydrodynamic pressure exerted by the head water, and p_e' is the hydrodynamic pressure exerted by the tailwater during an earthquake.

In a gravity dam, stability is secured by making it of such a size and shape that it will resist overturning, sliding and crushing at the toe. The dam will not overturn provided that the moment around the turning point, caused by the water pressure is smaller than the moment caused by the weight of the dam. This is the case if the resultant force of water pressure and weight falls within the base of the dam. However, in order to prevent tensile stress at the upstream face and excessive compressive stress at the downstream face, the dam cross section is usually designed so that the resultant falls within the middle at all elevations of the cross section (the core). For this type of dam, impervious foundations with high bearing strength are essential.

Stability Analysis

The stability of a gravity dam can be approximately and easily analyzed by two dimensional gravity method and three dimensional methods such as slab analogy method, trial load twist method, Finite Element Method (FEM), or by experimental studies on models. In this study, the stability of the dam will be analyzed by two dimensional gravity method and Finite Element Method (FEM), using ANSYS 5.4.

The steps involved in the analysis of gravity dam by analytical two dimensional gravity method are: 1) consider unit length of the dam; 2) work out the magnitude and directions of all the vertical and horizontal forces acting on the dam and their algebraic sum;

3) determine the moment arm of all these forces about the toe; 4) determine the moments of all these about the toe and find out the algebraic sum of all those moments; 5) find out the location of the resultant force by determining its distance from the toe; 6) find out the eccentricity of the resultant, which must be less than B/6 in order to ensure that no tension is developed anywhere in the dam; 7) determine the vertical stresses at the toe and heel; 8) determine the maximum normal stresses at the toe and heel; and 9) determine the factor of safety against overturning, sliding and crushing. The assumptions made in the two dimensional gravity method are: 1) the dam is considered to be composed of a number of cantilevers, each of which is 1 m thick and acts independent of the other; 2) no loads are transferred to the abutments by beam action; 3) the foundation and the dam behave as a single unit; 4) the materials in the dam body and foundation are isotropic and homogeneous; and 5) the stresses developed in the dam and foundation are within elastic limits and no movement of the foundation is caused due to transference of loads.

The finite element method (FEM) is a numerical method for determining responses (deformation, strain, stress, etc.) of a body under external loads. The finite-element method (FEM) uses a concept of "piecewise approximation". In theory, the elements can be of different shapes and sizes. Until developing FEM it was almost difficult to calculate point to point responses (approximately) of a body of any geometric shape and any complex type of loading conditions. In this method, the entire dam body is divided by using equivalent system of small triangular element for obtaining responses within and boundary (node) of the element. This method determines first the global deformations at the nodes of the element then determines successively other responses such as strains, stresses, etc. Nowadays, this method is available as a commercial FE programming or software (ABAQUS, ANSYS, STADD.pro, SAP, etc.) for solving large problems. Most of the FE programming used for either general purpose or special purpose follows the same basic procedure of finite element method.

RESULTS AND DISCUSSION

The section of gravity dam should be chosen in such a way that it is the most economic section and satisfies all the conditions and requirements of stability. Hence, after a design section of the dam has been arrived at, the stability analysis for the dam must be carried out. In this study, water pressure due to maximum pool level, extreme uplift pressure without any reduction due to drainage, tailwater pressure up to 6 m height (reservoir full case), and earthquake forces are considered in the design of gravity dam. The earthquake forces can be taken as equivalent to 0.15 g for horizontal forces and 0.10 g for vertical forces, considering the earthquake might occur in seismic zone 2 of Bangladesh. The uplift is considered as equal to the hydrostatic pressure at either ends. The unit weights of concrete and water are assumed to be 24.00 kN/m^3 and 9.81 kN/m^3, respectively. Assuming allowable concrete compressive strength of 3000 kN/m^2, the limiting height of low gravity dam is calculated to be 90 m. Thus, in order to consider the dam as high dam, the preliminary height of the high dam is increased to 95 m. A freeboard of 5 m equals to about 4% - 5% of the dam height is considered. Finally, the total height of the dam including freeboard is considered to be 100 m for stability and stress analyses. The u/s face is flattened to keep developed tension at the toe within allowable limit. For stability and stress analyses, the resulting high concrete gravity dam with design external forces is shown in Figure 2.

The stability analyses are performed for horizontal earthquake intensities varying from 0.10 g - 0.30 g with 0.05 g increment, keeping the vertical earthquake intensity constant at 0.10 g and unchanging other loads which may likely to occur in Bangladesh.

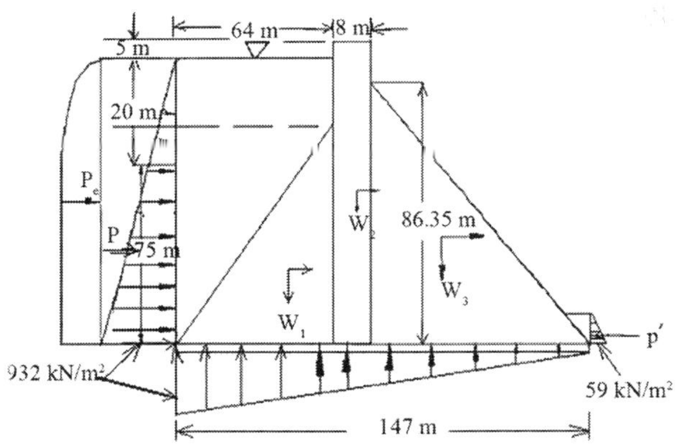

Figure 2: High concrete gravity dam with external forces.

For example, the stability and stress analyses are described in the following paragraph in the case of 0.15 g horizontal earthquake intensity and other constant design loads (i.e., the main case for which the dam to be constructed in seismic zone 2 of Bangladesh).

The total vertical and horizontal forces are calculated and the moments of all the forces about toe of the dam are taken. The value of eccentricity is found to be 20.04 m, which is less than B/6, i.e., 24.5 m, and hence, no tension will be developed in any portion of the dam. The vertical stress p_v at toe of the dam is found to be 1267.36 kN/m^2, which is less than allowable concrete compressive stress of 3000 kN/m^2; and 126.92 kN/m^2 at heel, which is less than the allowable concrete tensile stress of 500 kN/m^2. The principal stress at toe s (toe) is found to be 2178.93 kN/m^2, which is less than 3000 kN/m^2; and –412.75 kN/m^2 at heel s (heel), which is less than 500 kN/m^2. Neglecting small hydrodynamic pressure exerted by the tailwater (Negligence is on the safer side), the shear stresses t at toe and heel are found to be 1049.53 kN/m^2 and 632.42 kN/m^2, respectively, which are less than 3000 kN/m^2. All the stresses calculated at toe and heal of the dam are found to satisfy the conditions of safe design. The factors of safety against overturning and sliding, and shear friction factor are found to be 1.53, 1.001, and 3.87, respectively, which are greater than 1.5, 1.0, and 3.0, respectively, indicating the design is safe against these failures. An average value of the frictional coefficient, m = 0.7 (0.65 – 0.75) is considered to calculate the factor of safety against sliding and shear friction factor. For the set of loading conditions shown in Figure 2 and for economic dam section, the u/s face of the dam can be flatten up to a slope of 64H:75V in order to make it safe, but it cannot be flatten more than 64H:75V slope because the principal stress at the heel will be more than the allowable limit of 500 kN/m^2. If it is not possible to provide safe design by flattening the u/s face, either the top width of the dam or the d/s slope of the dam needs to be enlarged, but from economic and stability points of view, it is not feasible to widen the top width of the dam. The results for varying horizontal earthquake intensities, keeping other loads unchanged, are summarized in Table 1.

A static finite element analysis was also carried out with the same dam section and loading conditions used in 2D gravity method to investigate the stresses and deformations under the expected design loads. To simulate the process of activity of concrete gravity dam under given loading conditions, the assumptions made in the finite element analysis are: 1) the dam materials are homogeneous and isotropic; 2) since the nature of loading on the dam makes the dam problem as plain strain problem, therefore, it is analyzed as a plane strain problem; 3) the modulus of elasticity and Poisson's ratio were taken as 8.3 GPa [the modulus of elasticity was calculated based on supplied empirical formula $57500\times(f'_c)^{1/2}$ [13] and it was found 8.3 GPa for f'_c = 3000 kN/m^2] and 0.18 (Poisson's ratio for plain concrete generally varies from 0.15 - 0.20), respectively; and 4) no vertical and horizontal displacements are assumed to be occurred due to applied loads.

The finite element analysis was carried out using general purpose finite element software (ANSYS 5.4) for linear elastic analysis of concrete gravity dam. ANSYS FE programming consists of preprocessor, solution, and post processor parts. The FE modeling of a body is carried out in preprocessing part and post processing procedure is used to generate and interpret results. Six-noded quadratic triangular plane strain solid elements were used to discretize the entire gravity dam to achieve more accurate results. All nodes of elements have two degrees of freedom. The integration points in the case of reduced integration are 3 for six-noded quadratic triangular plane strain solid element.

The entire gravity dam was divided as three regular shaped parts for calculating self-weight of the dam. The self-weight of each part was assumed to act at its own center of gravity. The other loads, such as, weight and pressure of water at the u/s and d/s sides of the dam, hydrodynamic pressures of water at the u/s and d/s sides and uplift pressure of water at the bottom of the dam were applied as surface loads on the element's side. The horizontal earthquake intensities were perturbed from 0.10 g - 0.30 g with 0.05 g increment, keeping other loads unchanged. The horizontal earthquake loading of different intensities and vertical earthquake loading were applied at the point of self-load of each part of the dam.

A post processing procedure was used to generate results directly from ANSYS 5.4 finite element program. This program determines displacements at global coordinates and strains and stresses at local –x and –y coordinate systems as well as it can automatically determine the principal stresses. For example, the contour plot of shear stress (t) in the case of horizontal earthquake intensity of 0.15 g is shown in Figure 3.

Table 1: Comparsion of stresses obtained from 2D Gravity and FE methods and factors of safety based on 2D gravity method for varying K_h

Value of k_h	Design Dam Description	ΣM_R (+) kN.m	ΣM_O (−) kN.m	ΣV kN	ΣH kN	p kN/m²		Principal Stresses (σ) kN/m²		Shear Stresses (τ) kN/m²		Factor of Safety Based on 2D Gravity Method		
						2D Gravity method	FEM	2D Gravity method	FEM	2D Gravity method	FEM	F.S.O.	F.S.S.	S.F.F.
0.10	Total ht = 100 m Top width = 8 m U/s length = 49.5 m D/s length = 72.5 m Total base width = 130 m	12516408	8314995	86642	61258	Toe = 1256.5 Heel = 126.8	Toe = −544.1 Heel = 1005.0	Toe = 2061.2 Heel = −203.2	Toe = −548.4 Heel = 5.5	Toe = 989.0 Heel = 500.0	Toe = 232.8 Heel = 1022.0	1.505	1.013	3.98
0.15	Total ht = 100 m Top width = 8 m U/s length = 64 m D/s length = 75 m Total base width = 147 m	15815144	10336572	102429	71680	Toe = 1267.4 Heel = 126.9	Toe = −408.5 Heel = 1339	Toe = 2178.9 Heel = −412.7	Toe = −579.4 Heel = 436.2	Toe = 1049.5 Heel = 632.4	Toe = 193.9 Heel = 1037.0	1.53	1.001	3.87
0.20	Total ht = 100 m Top width = 8 m U/s length = 80 m D/s length = 95 m Total base width = 183 m	24317116	15205528	125319	87258	Toe = 1106.7 Heel = 262.9	Toe = −172.1 Heel = 1025	Toe = 2374.9 Heel = −412.1	Toe = −365.3 Heel = 36.5	Toe = 1152.7 Heel = 632.7	Toe = 163.4 Heel = 8.7	1.60	1.005	3.94
0.25	Total ht = 100 m Top width = 8 m U/s length = 100 m D/s length = 130 m Total base width = 238 m	40885835	24518064	158200	110780	Toe = 926.1 Heel = 403.8	Toe = −69.9 Heel = 742.8	Toe = 2891.3 Heel = −387.8	Toe = −196.1 Heel = 597.0	Toe = 1305.4 Heel = 593.8	Toe = 70.7 Heel = 269.4	1.67	1.000	4.01
0.30	Total ht. = 100 m Top width = 8 m U/s length = 136 m D/s length = 220 m Total base width = 364 m	95035252	54375374	229410	166462	Toe = 679.2 Heel = 580.8	Toe = −49.5 Heel = 222.6	Toe = 4709.0 Heel = −311.9	Toe = −35.9 Heel = 24.1	Toe = 1581.5 Heel = 492.3	Toe = 14.0 Heel = 339.1	1.75	1.001	4.18

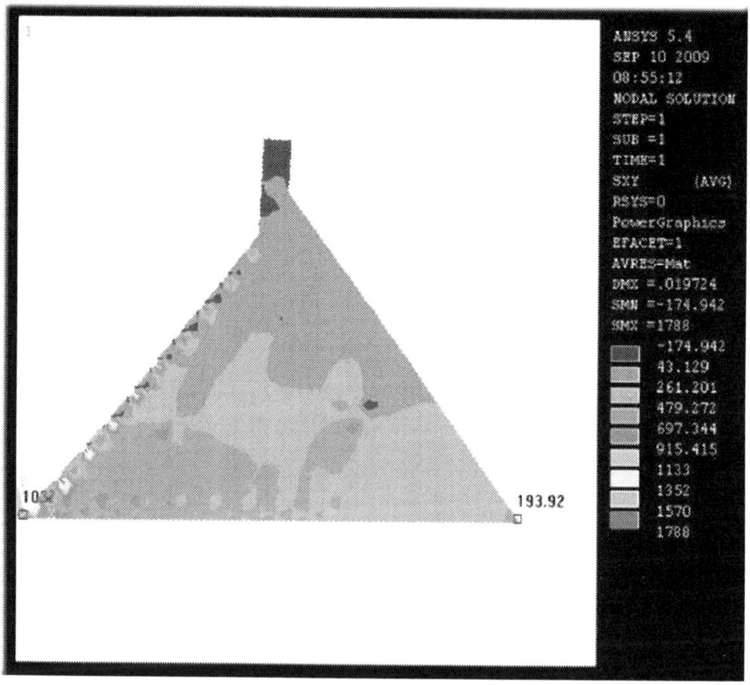

Figure 3: Contour plot of shear stress (t).

The variation of shear stress in the dam body can also be found with different color codes (Figure 3). The vertical, principal and shear stresses obtained using finite element analysis for varying horizontal earthquake intensities from 0.10 g - 0.30 g with 0.05g increment were compared with those obtained from the analytical 2D gravity method as shown in Table 1. Table 1 shows that for each intensity of horizontal earthquake acceleration the vertical stresses p_v obtained from ANSYS 5.4 vary significantly from the stresses obtained using analytical 2D gravity method; because the vertical stresses in the case of ANSYS 5.4 were found always negative at the toe, whereas no negative values were found using 2D gravity method. However, the vertical stresses found at the toe and heel of the dam for 0.10 g - 0.30 g using these two methods respectively are within the concrete allowable compressive stress. Similarly, the principal and shear stresses at the toe and heel obtained from ANSYS are also found to vary significantly with those obtained from the 2D gravity method (Table 1). The principal stresses at the toe of the dam were found to exceed the allowable concrete tensile stress slightly for 0.10 g - 0.15 g only. The principal stresses can be made safe by further flattening the u/s side of the dam, i.e., by further increasing the self-weight of the dam. The shear stresses obtained using these two methods for 0.10 g - 0.30 g are found to satisfy the allowable concrete compressive stress successfully and ANSYS results are found to vary in similar fashion with those obtained using 2D gravity method, specifically at the toe of the dam (Table 1).

The vertical, principal and shear stresses found using ANSYS are more conservative than those obtained using 2D gravity method, except the principal stresses at the toe of the gravity dam for 0.10 g - 0.15 g. This significant variation of results could be happened, because in analytical procedure all surface loads are assumed to act as concentrated load at its own center of gravity and the determination of stresses at the toe and heel is based on average values, while in FE analysis it is the determination of point to point stresses of the entire dam under actual loading conditions and thus the FE analysis can be regarded much more reliable than any other methods. Only the principal stresses found at the toe of the dam using FE analyses for 0.10 g - 0.15 g and using 2D gravity method for 0.30 g were found to exceed the allowable concrete stress throughout the analyses. The principal stresses found at the toe of the dam using FE analyses for 0.10 g - 0.15 g can be made safer by just flattening the u/s face of the dam, but it might not be possible to make it safe in the case of 2D gravity method for 0.30 g. Moreover, it was very difficult to achieve the required factor of safety against sliding, specifically for 0.30 g horizontal earthquake intensity. Although, it suggests apparently that smaller dam section may be enough for stress analyses using FEM (Table 1), it would not possible to achieve the required factors of safety with smaller dam section. Thus, further dynamic analyses using this/other FE methods may be carried out to obtain more reliable information to make necessary comments on the stability of the dam.

The stabilizing moments are determined using analytical 2D method dealing with the U.S.B.R. recommended initial dam section and found to decrease significantly with the increment of horizontal earthquake intensity, indicating endanger to the dam stability, thus larger dam section is designed to increase the stabilizing moments and to make it safe against failure. The factors of safety against overturning and sliding, and shear friction factor were determined to examine whether these factors are safe against these failures. It is observed that the factor of safety against sliding is satisfied at last than other factors of safety, resulting huge dam section to make it safe against sliding and making impossible to provide at least 1:1 ratio of dam base width to dam height, using analytical 2D gravity method (Table 1).

Based on the 2D gravity method, the volume of concrete required per meter length of the gravity dam to make it safe against failure for varying horizontal earthquake intensities and unvarying other loads is found to increase exponentially as shown in Figure 4. During factor of safety determination, it is observed that the factor of safety against sliding is satisfied much later than other factors of safety, allocating huge dam section to make it safe against sliding. Therefore, sufficient base width, silt pressure, adequate strong rock foundation, drainage gallery to reduce uplift pressure, and construction joints to increase frictional resistance need to be ascertained to improve factor of safety against sliding as well as provisions should be made for shrinkage and temperature stresses. The shrinkage and temperature stresses will cause the concrete to crack unless remedial measures are undertaken [7]. These stresses may be avoided by using minimum amount of concrete, low lift of concrete (generally 1.5 m), construction joints, special low heat cement, and cooling aggregates and concrete surface, etc.

Initially, the section of the dam is selected arbitrarily based on the U.S.B.R. recommendations. This arbitrary selection of the gravity dam profile is much more complex than actual profile, because in actual profile at least the height and top width of the dam are fixed based on the requirements. Thus, the selection technique of the dam profile as well as the stability analysis procedures can be applied to the actual dam design. The detailed factors of safety and stresses analyses are described and comparison between the two methods has been made (Table 1). Finally, it may be concluded that it would not

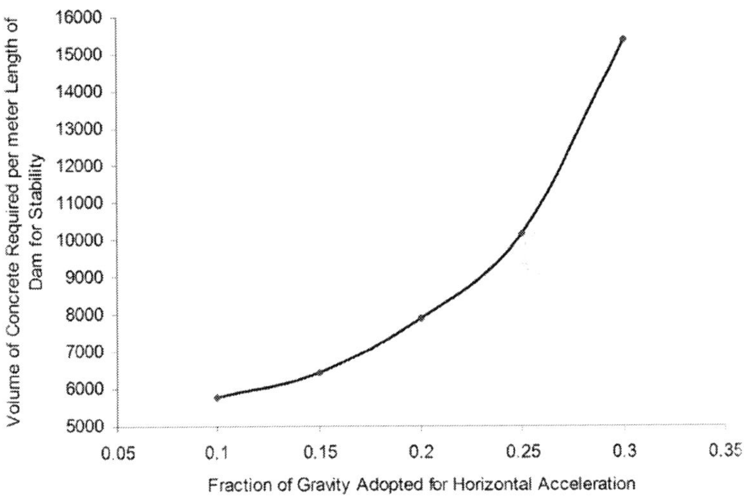

Figure 4: Volume of concrete (m^3) required per meter length of gravity dam for stability against varying horizontal earthquake intensity.

be feasible to construct a concrete gravity dam in seismic zone 2 of Bangladesh for k_h values greater than 0.30 without changing other loads and/or dimension of the dam, and without keeping provision for drainage gallery to reduce the uplift pressure significantly at the base of the dam.

CONCLUSIONS

The section of gravity dam should be chosen in such a way that it is the most economic section and satisfies all the conditions and requirements of stability. The preliminary dam section is selected based on the U.S.B.R. recommendations, and the stability and stress conditions of the high concrete gravity dam are approximated and analyzed for varying horizontal earthquake intensity and unvarying other loads, using two-dimensional gravity method and finite element method. The horizontal earthquake intensities are perturbed from 0.10 g - 0.30 g with 0.05g increment, keeping other loads unchanged, to calculate the total horizontal and vertical forces and moments at the toe of the gravity dam, and to examine the stability and stress conditions of the dam using the two methods. Dealing with the U.S.B.R. recommended initial dam section, the stabilizing moments are found to decrease significantly with the increment of horizontal earthquake intensity, indicating endanger to the dam stability, thus larger dam section is designed to increase the stabilizing moments and to make it safe against failure. The results of the horizontal earthquake intensity perturbbation suggest that the stability of the gravity dam endangers with the increment of horizontal earthquake intensities unless the dam section is enlarged significantly. The FE analyses were carried out using the same dam section and loading conditions used in 2D gravity method. The vertical, principal and shear stresses using ANSYS 5.4 are found more conservative than those obtained using 2D gravity method, except the principal stresses at the toe of the gravity dam for 0.10 g - 0.15 g. Only the principal stresses found at the toe of the dam using FE analyses for 0.10 g - 0.15 g and using 2D gravity method for 0.30 g, were found to exceed the allowable concrete stress throughout the analyses. The principal stresses found at the toe of the dam using FE analyses for 0.10 g - 0.15 g can be made safer by flattening the u/s face of the dam, but it might not be possible to make it safe in the case of 2D gravity method for horizontal earthquake intensity of 0.30 g. Moreover, it was very difficult to achieve the required factor of safety against sliding, specifically for 0.30 g horizontal earthquake intensity. Although, one can feel apparently that smaller dam section is required for stress analyses using FEM, it would not possible to achieve the required factors of safety with smaller dam section. It is also observed that the factor of safety against sliding is satisfied at last than other factors of safety, resulting huge dam section to make it safe against sliding. Therefore, sufficient base width, adequate strong rock foundation, drainage gallery to reduce uplift pressure, silt pressure, and construction joints need to be ascertained to improve factor of safety against sliding. Finally, it can be concluded that it would not be feasible to construct a concrete gravity dam for k_h values greater than

0.3 without changing other loads and or dimension of the dam and keeping provision for drainage gallery to reduce the uplift pressure significantly.

ACKNOWLEDGEMENTS

This research was supported by funds provided by the Department of Civil Engineering, Chittagong University of Engineering and Technology (CUET), Bangladesh. The authors gratefully acknowledge the technical support given by Dr. Sobri Harun, Associate Professor, Dept. of Hydraulics and Hydrology, UTM, Malaysia.

REFERENCES

1. M. F. Kennard, C. C. Owens and R. A. Reader, "An Engineering Guide to the Safety of Concrete and Masonry Dams in the UK," Report 148, CIRIA, 1995.
2. N. Smith, "A History of Dams," Peter Davies, London, 1971.
3. A. Vogel, "The Historical Development of the Gravity Dam," in: H. Fahlbusch, ed., Historical Dams, International Commission on Irrigation and Drainage, New Delhi, 2001, pp. 61-70.
4. A. K. Biswas and S. Chatterju, "Dam Disasters: An Assessment," J. Eng. Inst. Can., Vol. 54, No. 3, 1971.
5. G. B. Baecher, M. E. Gregory and R. de Neufville, "Risk of Dam Failure in Benefit-Cost Analysis," Water Resources Research, Vol. 16, No. 3, 1980, pp. 449-456.
6. Bureau of Reclamation, "U.S. Department of the Interior Design of Small Dams," 1977.
7. S. K. Garg, "Irrigation Engineering and Hydraulic Structures," 16th Edition, Khanna Publishers, Delhi, 2002, pp. 960-1020.
8. "Bangladesh National Building Code," 1st Edition, Housing and Building Research, Institute and Bangladesh Standards and Testing Institute Publishers, Dhaka, 1993.
9. C. N. Zangar, "Hydrodynamic Pressures on Dams due to Horizontal Earthquake Effects," Engineering Monograph, No. 11, Bureau of Reclamation, 1952.

10. C. V. Davis, "Handbook of Applied Hydraulics," McGrawHill, Basingstoke, 1969.
11. G. Oberti, "The Interconnection between Concrete Dam and Foundations," ICOLD, XVII Congress, Q. 66, R. 42, Vienna, 1991.
12. L. R. Volpe, S. C. Ahlgren and E. R. Goodman, "Selection of Engineering Properties for Geologically Variable Foundations," ICOLD, XVII Congress, Q. 66, R. 59, Vienna, 1991.
13. G. Winter, L. C. Urquhart, C. E. O'Rourke and A. H. Nilson, "Design of Concrete Structures," 7th Edition, McGraw-Hill Inc., Boston, 1964, p. 660.

Chapter 11

Use of Stone Dust in the Design of High Performance Concrete

Joffrey Cheruiyot, Sylvester Ochieng Abuodha, and Charles Kabubo

SMARTEC, Jomo Kenyatta University of Agriculture and Technology (JKUAT), Nairobi, Kenya

ABSTRACT

This research evaluated the suitability of stone dust in the design and production of High Performance Concrete (HPC). HPC mix was designed, tested, costed and a comparison of concrete classes used in the market (Class 25, 30 and 35) done using Cost Benefit Analysis (CBA). The cost benefit was analyzed using Internal Rate of Return (IRR) and Net Present Value (NPV). Laboratory tests established the properties concrete obtained from the design mix. Compressive strength, slump, and modulus of elasticity were tested and analyzed. Structural analysis using BS 8110 was done for a 10 storey office building to establish the structural member sizes. Members obtained from concrete Classes 25, 30, 35 and the new compressive strengths from HPC (Class 80) were obtained and compared. Analysis was done for structural members' sizes and area freed as a result of designing with HPC as well as the steel reinforcement used. To justify the initial cost of HPC if adopted, the Cost Benefit Analysis (CBA) was used to estimate increased costs versus income resulting from increased let table space created. The minimum class of concrete used in design was limited to Class 25 N/mm^2. The research shows that it is possible to manufacture high strength concrete using locally available stone dust. The stone dust sampled from Mlolongo quarries achieved a characteristic strength of 86.7 N/mm^2 at a water cement ratio of 0.32. With the results structural analysis of a 10 storey office structures with columns spaced at 8 meters center to center was designed using the four classes and results compared. There was a reduction of columns from 1.2 m wide to 0.65 m wide (over 45%) when concrete class changes from Class 25 to Class 80 creating over 3% of the total space area per floor. Cost benefit analysis using Net Present Value (NPV) and Internal Rate of Return (IRR) presented business case for the use of HPC. With Class 80, the IRR was at 3% and NPV being 8% of the total initial investment. The steel reinforcement increased by 8.64% using Class 30, 11.68% using Class 35 and reduced by 8.37% at Class 80. Further analysis needs to be done to understand the trend of steel reinforcement keeping all the member sizes the same. In this study the member sizes were optimized based on the steel reinforcement and serviceability. This paper provides useful information to design Engineers and Architects and inform future design of multi storey structures.

INTRODUCTION

More than 40 years have passed since High Performance Concrete was realized; high-strength concrete itself has been enhanced in terms of performance and placeability through a lot of useful developments to the stage that now many applications have been done in North America, Europe, Japan and other countries (Kuwai, T, 2001) [1] .

Even though applications of HPC have been greatly used in Japan, Europe and United States of America, little has been done here in Kenya. Research findings, and cost benefits analysis done for the 100 year design life using HPC will provide the engineering practitioners the necessary information for their design. Use of stone dust will also enhance environmental conservation particularly where river sand use will be reduced.

Objectives

- Evaluate the suitability of stone dust (quarry waste) in production of High Performance Concrete (HPC).
- To compare the cost of HPC (Class 80) production against costs of traditional Class 25, 30, and 35 MPa.

THE LITERATURE REVIEW

High performance concrete (HPC) is defined as concrete whose strength and durability is greater than those of "normally" obtained concrete (Addis, 2001) [2] . According to United States of America Strategic Highway Research Program on high performance concrete, HPC is defined as having 4 hour compressive strength of greater than or equal to 17.2 Megapscals (Mpa), 24 hour compressive strength of greater than 34.5 Mpa and 28 day compressive strength of greater than 68.9 Mpa (Zia et al., 1991) [3] .

High Strength Concrete (HSC) is an example of HPC and according to Japanese Standards of Civil Engineers (JSCE), high strength concrete is any mix with strengths of between 50.0 Mpa to 70.5 Mpa. Although more expensive, per cubic meter, than convectional concrete, it can carry a given load more economically (Rusell et al. 1997) [4] .

More than 40 years have passed since High Performance Concrete was realized; high-strength concrete itself has been enhanced in terms of performance and placeability through a lot of useful developments to the stage that now many applications have been done in North America, Europe, Japan and other countries (Kuwai T., 2001) [1] .

Since the introduction of concrete with a compressive strength of 62 MPa in columns, shear walls and transfer girders of the Water

Tower Place in Chicago in 1975, many applications of HSC in projects, ranging from transmission poles to the tall buildings (KLCC Twin Tower in Kuala Lumpur, Malaysia) with concrete strength reaching up to 131 MPa in the Union Square building in Seattle, Washington have been reported.

In the near future, the high-strength concrete will be prevailing and increasing in use all over the world. The current trend of construction development which has increased over time, there is need to develop economic and environmental friendly construction materials.

There is considerable work done so far on high performance concrete internationally and several research done on stone dust (sometimes referred to as quarry dust) here in Kenya. Stone dust is classified as material obtained from stone crushing process, and due to its fines rarely used in concrete design because of high water cement ratio requirement (Adis, 1991) [5] .

Kiliswa (2012), [6] studied the effect of quarry dust on the strengths and permeability of concrete (Class 20 and 25). From his results he concluded that concrete designed with quarry dust as a substitute of river sand had plastic and hardened properties equal to or greater than traditional mixes (aggregate, river sand and cement).

Through the use of high-strength concrete floor, areas can be increased because it allows the sectional area of building columns to be reduced (Jinnai, H. et al., 2005) [7] . Although more expensive, per cubic meter, than convectional concrete, it can carry a given load more economically (Rusell et al. 1997) [4] .

Whereas there are documented application in Japan, Europe and America, Kenyan construction industry has not embraced the technology. This research will stimulate the discussions around percentage savings in lettable space in highrise buildings when HPC is used in the design. With the strengths of concrete achieved, a typical high-rise building will be designed using Class 25, and 35 and HPC design strength achieved. Structural member sizes will be compared and a cost benefit analysis done for 100 design year period.

THE METHODOLOGY

Materials Selection

Stone dust was collected from Athi River crushing plant for testing after stock pile sampling. Ordinary Portland cement (CEM 42.5) and admixture Sika® ViscoCrete®-HE, a water reducer and super plasticizer obtained from the local dealers were used in the design for concrete manufacture.

Design and Manufacture and Tests of Concrete (HPC)

Batching was done by mass for cement and stone dust (American Concrete Institute, 1997) [8] while admixtures in liquid were measured in liters using the conversion factor from the Sika® Visco Crete®-HE technical data sheet provided. Mixing was done in a forced action mixer. The use of free fall mixers and hand mixing is not permitted (Addis, 2001) [2] .

Sika® ViscoCrete®-HE a high range water reducing and superplasticizing admixture was used to increase plasticity of fresh concrete. 0.6% - 0.8% of the admixture by weight of cement was used.

Mixing Method

Concrete were mixed in a tilting mixer (laboratory type). The following mixing sequence was adopted: stone dust and cement were mixed for 2 minutes; water added during mixing and continuous mix for two more minutes; mixing was stopped for one minute; admixture added to the mix and 3 minutes more minutes mixing.

The methodology adopted was used by Aminul, I.L., in his research Mix Design of High Performance Concrete, National Institute of Technology Silchar, India, 2008 [9] . The methodology was also adopted by Addis. B.J in paper—A method of proportioning trial mixes for high strength concrete published by ACI in 1990 [10] .

Slump test, Compressive Strength, and Modulus of Elasticity were tested and results analyzed to get the structural properties of obtained HPC.

11 cubes were cast per mix; 1 for one day test, 2 for seven day test, 4 for twenty eight day test and 4 for one year test (Concrete strength to be checked in a year's time).

Structural Design of 10 Storey Structure

A typical high rise structure (10 floors) was designed using British Standards for reinforced concrete design (BS 8110). Structural member sizes determined for several grades of concrete (Class 25, 30, 35 and 80, the new HPC strength obtained). Ultimate Limit State (ULS) design philosophy was used where serviceability as well as ultimate limit states will be checked. The SLS for deflection considered the performance of the structure under working Loads. The design ensures that the structure, or part of it under consideration, does not deflect excessively causing unsightly cracking and loss of durability.

Figure 1 shows the general layout of the office structure designed, the architect was specific with the column spans at 8 meters center to center. Thus the structure was designed with ribbed slab, 4 meter ribs supported by a slab beam simply supported from the main frame moving across.

Cost Benefit Analysis

The cost of production of HPC (Class 80) was calculated based on optimum design mix obtained from the laboratory tests. This cost was compared with the costs of Class 25, 30, and 35 as available in the market. Further Cost Benefit Analysis (CBA) was undertaken to evaluate the benefit accrued as a result of increased lettable space over the investment incurred. CBA determines the soundness of the investment/ decision (justification/ feasibility). It provides a basis for comparing concrete classes.

Total costs comparisons expected for each option against the total expected benefits, to see whether the benefits outweigh the costs, and by how much.

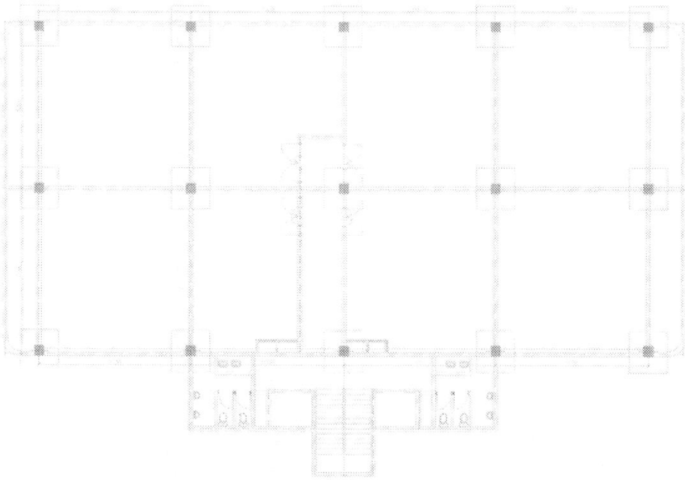

Figure 1: Layout of the structure.

RESEARCH RESULTS AND DISCUSSION

Data Presentation

Material Properties

Crushed stones was sampled from Mlolongo Quarry and taken to the laboratory for tests. Grading, Finess Modulus (FM), Specific Gravity (SG) and Water Absorption (WA) was tested.

Table 1 and Figure 2 summarizes the material properties of the material used for test.

The material was graded from 75 μm to 10 mm particle size. The samples used had water absorption ratio of 5.8%, relative density of 2.389 and Finess Modulus of 3.0.

Design Mix and Results

5 trial runs were possible from a water cement ratio of 0.42, 0.38, 0.35, 0.32, and 0.29 beyond which the mix became so dry. The range of water cement ratio for high strength concrete according to Russel et al., 19,774 ranges from 0.42 to 0.25 using stone dust in South Africa.

From the results, maximum strength of 92.7 N/mm^2 (W/C of 0.32) was obtained and a minimum of 62 N/mm^2 (W/C of 0.35).

Factoring in standard deviation; fk = fc + (1.4 × Standard Deviation), limited to 3.5 based on the laboratory results.

Therefore 92.7 = fc + 1.4 × 3.5, hence fc = 87.8 N/mm^2.

To analyze cost benefits of high strength concrete Class 80 were be compared to the common Class 35, 30 and 25 N/mm^2.

Table 2 shows the water cement ratio used in design of concrete mixes and the attained compressive strength of concrete cubes at 28 day tests.

The compressive strength results are summarized in Figure 3 and follows the normal trend with increased strength with reduced water cemet ratio. The maximum strength was achieved at a water cement ratio of 0.32.

Table 3 summarizes the mix designs and the results achieved at 1 day compressive strength for 150 mm cubes, 7 day test and 28 day tests. It also shows the slump achieved and the mix design details.

Modulus of Elasticity

The Average Modulus of elasticity obtained from the results was 49.4 GPa, well above the normal concrete range of 20 - 30 GPa. Figure 4 shows stress strain curve for the HPC 150mm by 150mm cubes tested using a compressive strength machine and a strain gauge. Figure 4 shows the stress in N/mm^2 plotted against strain and Modulus of Elasticity calculated.

Discussion

From the results, the optimum mix design is obtained at a water cement ratio of 0.32 with strength of 92.7 KN/mm^2. Factoring in standard deviation; fk = fc + (1.4 × Standard Deviation), limited to 3.5 based on the la- boratory results the characteristic 87.8 N/mm^2 Compressive Strength is obtained. For design and cost benefits analysis Class 80 was subsequently used.

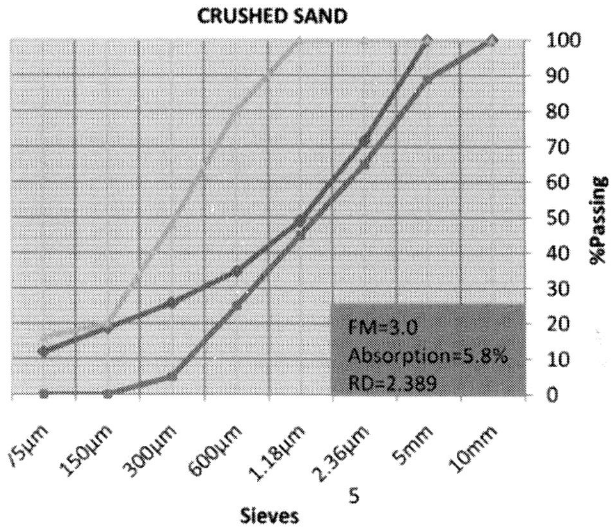

Figure 2: Material properties for stone dust used for concrete design. The green and red lines represent the upper limit and lower limits as per BS specification on crushed sand.

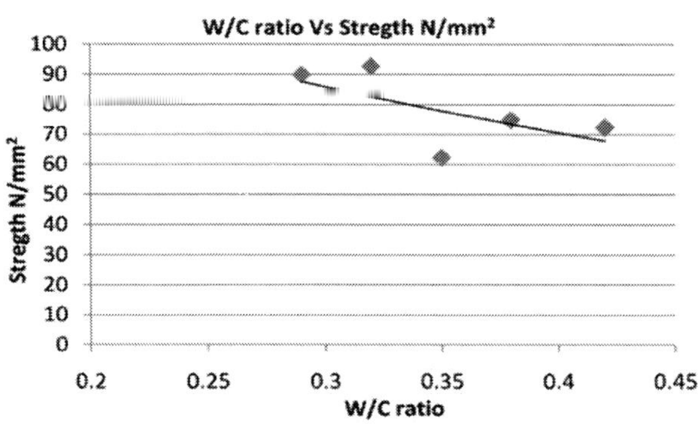

Figure 3: Water cement ratio versus compressive strength obtained.

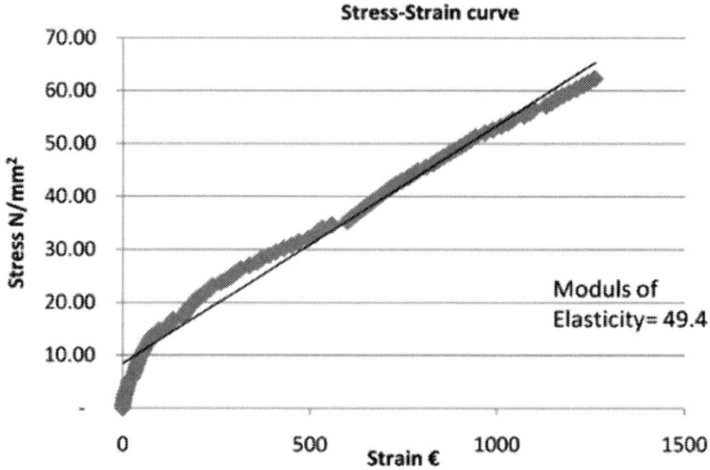

Figure 4: Stress versus strain curve.

Table 1: Crushed stone dust grading results

INDIVIDUAL GRADING CURVES			
% Passing by Weight			
Sieve sizes	**CRUSHED SAND Achieved**	**LL**	**UL**
10 mm	100	100	100
5 mm	100.0	89	100
2.36 μm	71.6	65	100
1.18 μm	49.0	45	100
600 μm	34.8	25	80
300 μm	25.8	5	48
150 μm	18.7	0	20
75 μm	12.0	0	16

Table 2: WC rato/Compressive strength obtained

	STRUCTURAL MEMBER SIZES					
#	Structural Member	Size (mm)	**Concrete Class (N/mm²)**			
			25	**30**	**35**	**80**
1	Internal Colums	b	1200	1000	900	650
		h	1200	1000	900	650
2	External Columns	b	1000	900	800	500
		h	1000	900	800	500
3	Internal Column bases	hf	400	400	400	400
4	External Column bases	hf	250	250	250	250
5	Beams (ribbed slab)-4 m long-320 No	b	125	125	125	125
		h	250	250	200	200
6	Slab beam-8 m long	b	500	500	500	500
		h	250	250	250	200
7	Secondary beams (supporting slab beam-8 m long-4 No	b	200	200	200	200
		h	450	450	450	300
8	Main frame Beam-8 m long-10 No	b	200	200	200	200
		h	600	600	450	425
9	External Beam-8 m long-12 NO	b	200	200	200	200
		h	450	450	350	250
10	Slab	hf	150	150	150	150

Table 3: Design mixes and laboratory results

FIRST BATCH DESIGN								
Cement	Cube ref	Age	Class/mix	Mix Details	Slump	Density	Strength	Average (N/mm²)
					mm	**Kg/m³**	**N/mm²**	
42.5	A	1	C70-MIX 3	CEMENT = 595	200	2187	27.5	27.5
	B	7		C-SAND = 1000		2200	60.1	59.0
	C	7		WATER = 250		2200	58.0	
	D	28		VISCO = 4.76		2230	74.5	73.5
	E	28		W/C RATIO = 0.42		2230	72.0	
42.5	A	1	C70-MIX 2	CEMENT = 595	180	2200	25.0	25.0
	B	7		C-SAND = 1318		2210	56.5	55.8
	C	7		WATER = 250		2230	55.0	
	D	28		VISCO = 3.57		2230	72.5	72.3
	E	28		W/C RATIO = 0.42		2240	72.0	

42.5	A	1	C70-MIX 4	CEMENT = 595	125	2195	24.0	24.0
	B	1		C-SAND = 1318		2200	24.0	
	C	7		WATER = 211		2170	47.0	47.5
	D	7		VISCO = 3.57		2170	48.0	
	E	28		W/C RATIO = 0.35		2160	62.0	62.3
	F	28				2210	62.5	

Structural Design

A typical structure was analyzed to obtain all the structural members. The analyzed building has ten storeys and a size of 38.6 m by 22.6 meters with floor spans of 8 m by 8 m. Since the floor spans are large, ribbed slab was designed for the 8 m by 8 m; the entire remaining floor spaced used a common 150 mm slab.

Structural members were obtained as outlined in Table4

The volume of concrete and steel reinforcement was also calculated from the members and it was established that concrete volume reduces by 29.9% when concrete strength is increased from Class 25 to Class 80.

Figure 5 shows volume of concrete in structural members and the total reinforcement required for the structure.

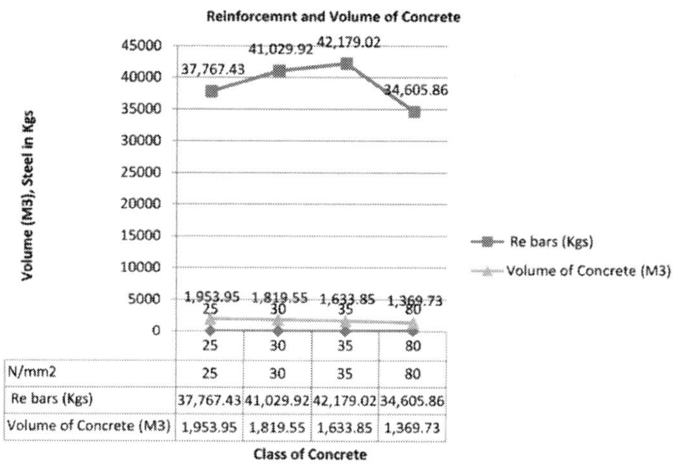

Figure 5: Comparison between volume of concrete and reinforcement steel.

Table 4: Structural member sizes obtained from the structural design

	STRUCTURAL MEMBER SIZES					
#	Structural Member	Size (mm)	Concrete Class (N/mm^2)			
			25	30	35	80

1	Internal Colums	b	1200	1000	900	650
		h	1200	1000	900	650
2	External Columns	b	1000	900	800	500
		h	1000	900	800	500
3	Internal Column bases	hf	400	400	400	400
4	External Column bases	hf	250	250	250	250
5	Beams (ribbed slab)-4 m long-320 No	b	125	125	125	125
		h	250	250	200	200
6	Slab beam-8 m long	b	500	500	500	500
		h	250	250	250	200
7	Secondary beams (supporting slab beam-8 m long-4 No	b	200	200	200	200
		h	450	450	450	300
8	Main frame Beam-8 m long-10 No	b	200	200	200	200
		h	600	600	450	425
9	External Beam-8 m long-12 NO	b	200	200	200	200
		h	450	450	350	250
10	Slab	hf	150	150	150	150

Figure 5 details the volume of concrete used in structural members when Class 25, 30, 35 and 60 is used as well as the weight of steel reinforcement required.

The volume of concrete reduces from 1953 when Class 25 is used to 1369 when Class 80 is used. The weight of steel reinforcement increases from Class 30 and Class 40 and drops significantly when Class 80 is used.

Floor area was calculated factoring in the area taken by the columns and it was established that additional 1.5% (13.4 m^2 of 874 m^2) of the floor area was added per floor when concrete strength was increased from Class 25 to Class 80. Table 5 summarizes freed space in the structure. The floor area will bring in additional income from the lease throughout the life cycle of the structure.

Cost Benefit Analysis

Internal rate of return (IRR) was calculated based on current concrete costs, cost of steel and projected earnings from the extra space obtained from reducing the structural members. The internal rate of return on an investment or project is the "annualized effective compounded return rate" or "rate of return" that makes the net present value (NPV as

$NPV = \sum_{t=1}^{t} \frac{Ct}{(1+t)^t} - Co$) of all cash flows (both positive and negative) from a particular investment equal to zero.

$$NPV = \sum_{t=1}^{t} \frac{Ct}{(1+t)^t} - Co$$

where C_t is the net cash in flow;

C_o is the initial investment;

r is the discount rate;

t is the time in years.

The IRR for Class 80 N/mm^2 was found to be 3% which reduces to 2% for Class 35% and 30% and 0% for Class 25.

Table 6 summarizes the NPV and IRR. The results shows that whereas the initial cost investments for Class 80 concrete is the highest, the net present value is lower and comparable with using Class 35. The internal rate of return for using Class 80 is 3% and thus show that it makes business sense to use Class 80 concrete.

Table 5: Area freed from use of smaller members and estimated returns

Concrete Class N/mm^2	Total Additional Area (m^2)	Benefits-Returns/m^2(KES)/ month
25	-	-
30	67	59,895.00
35	96	86,355.00
80	134	120,375.00

*Class 25 is the minimum class of concrete and the base used for analysis; *The cost of hiring space estimated at KES 850/m^2 per month

(Average market price in the CBD-Nairobi).

Table 6: Cost benefit analysis presentation

Concrete Class N/mm^2.	**Current Price for Construction (KES)**	**Net Present Value (NPV)**	**Internal Rate of Return (IRR)**
25	31,887,391	31,887,391	0%
30	33,718,185	32,412,497	2%
35	37,738,483	35,855,978	2%
80	38,114,759	35,490,631	3%

Summary

Quarry dust obtained from Mlolongo quarry graded between 75 μm to 19 mm with a relative density of 2.389 Kg/m^2 make good high strength concrete of over 80 N/mm^2 at a water cement ratio of 0.32. The concrete manufactured used Sika® ViscoCrete®-HE admixture which is a plasticizer and water reducer to minimize the water requirement and achieve the required slump.

Stone dust is much cheaper than both course aggregate and sand and therefore minimize the cost of concrete which has high cement volume per tonne (780 kg/m^3 from the mix design).

Design of structures using high strength concrete reduces significantly the structural member sizes, steel reinforcement and increase the available usable space in the structure. The limitations for reduced structural members are modulus of elasticity particularly for slender columns and shear reinforcements for foundation footing and heavily loaded beams.

There is significant return on investment when high strength of concrete is used for structural members (in the case of construction of 10 storey structure, the IRR was 3%). Use of HPC present additional advantages; in the case where the client has a small space and needs bigger floor area or in the case where the floor areas have big spans (architectural requirements of spaced columns).

CONCLUSIONS

From the research the following conclusion can be made:

Ÿ Manufacture High strength concrete (HPC) using locally available stone dust is possible with strengths of over 80 kN/mm^2 with a Modulus of Elasticity of 49.4 Gpa achieved using stone dust obtained from Mlolongo Quarries.

Ÿ Use of High strength concrete significantly reduces structural members (columns sizes, beam depths). The sizes are however limited by modulus of elasticity particularly for slender columns and shear reinforcement for heavily loaded columns and foundation pads.

Ÿ There is also significant reduction in the total weight of reinforcement steel when high strength concrete is used.

Ÿ The benefits accrued from letting extra space created from the smaller column sizes are significant and present a business case when life cycle costs of the structure are considered. This is important for clients who have smaller development space and want to maximize it or for structure that have bigger floor spans.

Recommendations

The construction industry needs to consider use of high strength concrete (HPC) in the current market as it will greatly improve the construction standards. There is enough justification for its use in high storey structures especially in rapidly congested areas of the cities where land has reduced.

Use of HPC will also help architects improve their designs particularly where they require big spaces or minimum columns at the centre of the structure.

Use of stone dust will also reduce river sand use which has in the recent past presented serious environmental impacts in the rivers where they are mined.

Areas of Further Study

As the laboratory work came to an end during this study, Bamburi Cement a leading manufacturing company in Kenya introduced

Ordinary Portland Cement Class 52.5. It will be important to see how the new product will enhance the concrete strengths.

There is need to comprehensively look at the relationship between reinforcements and class of concrete keeping all the structural members constant. In this study, where there was significant reduction of steel, like in the case of beams when compressive reinforcement was not required, the structural member was reduced for it to have more economic sense.

It will also be interesting to study unreinforced concrete beams using HPC. Modulus of rapture will be obtained and compared using the existing formulae.

REFERENCES

1. Kuwai, T. (2001) State-of-the-Art Report on High-Strength Concrete in Japan Recent Developments and Applications. 105-8007, 88.
2. Addis, B. and Owens. G. (2001) Fulton's Concrete Technology. 8th Edition. Cement & Concrete Institute, Midrand, South Africa.
3. Zia, P., Leming, M.L., Ahmad, S.H., Shemel, J.J., Elliot, R.P. and Naaman, A.E. (1991) Mechanical Behavior of High Strength Concrete, Strategic Highway Research Program, High Performance Concretes, a State of the Art report, SHRP National Research Council, Washington DC.
4. Russel, H. and Moreno, J. (1977) State-of-the-Art Report on High Strength Concrete, Chicago Committee on High-Rise Buildings.
5. Addis, B.J. (1991) Properties of High-Strength Concrete Concrete Made with South African Materials. Ph.D. Thesis, University of Witwatersand, Jonanesburg.
6. Kiliswa, M. (2011) Effect of Quarry Dust on the Strengths and Permeability of Concrete. Msc Thesis, University of Nairobi, Kenya.
7. Jinnai, H. (2005) Development and Construction Record on High Strength Concrete with the Compressive Strength Exceeding 150MPa, ACI SP-228, Seventh International Symposium on the Utilization of High-Strength/High-Performance Concrete, 1045-1062.

8. American Concrete Institute 211.1-9191997, Standard Practice for Selecting Proportions for normal, heavyweight, and mass concrete, Farmington Hills, Michigan: American Concrete Institute, 1997.

9. Aminul, I.L. (2008) Mix Design of High Performance Concrete, National Institute of Technology Silchar-788010, India.

10. Addis. B.J. and Alexander, M.G. (1990) A Method of Proportioning Trial Mixes for High Strength Concrete, High Strenth Concrete, Second International Symposium, Berkeley, California, 227-278 (ACI special Publication, SP-121).

Citations

CHAPTER 1

Juntao Chen, Yang Yang, Chao Ye, Ying Yang, and Ming Xiao, "Three-Dimensional Numerical Analysis of Compound Lining in Complex Underground Surge-Shaft Structure," Mathematical Problems in Engineering, Article ID 387379, in press.

CHAPTER 2

Seung-Jun Kwon, Byung Jae Lee, and Yun Yong Kim, "Concrete Mix Design for Service Life of RC Structures under Carbonation Using Genetic Algorithm," Advances in Materials Science and Engineering, vol. 2014, Article ID 653753, 13 pages, 2014. doi:10.1155/2014/653753.

CHAPTER 3

Chao Su, Zhenxue Zhu, Yangyang Zhang, and Niantang Jiang, "Optimization Design and Application of Underground Reinforced Concrete Bifurcation Pipe," Mathematical Problems in Engineering, Article ID 948584, in press.

CHAPTER 4

Lukasz Sadowski, "Methodology for Assessing the Probability of Corrosion in Concrete Structures on the Basis of Half-Cell Potential and Concrete Resistivity Measurements," The Scientific World Journal, vol. 2013, Article ID 714501, 8 pages, 2013. doi:10.1155/2013/714501.

CHAPTER 5

Q. Yu, H. Shimada, T. Sasaoka and K. Matsui, "Impact of Underground Mining on Shaft Lining and Aquifer in Eastern Chin," *Open Journal of Geology*, Vol. 2 No. 3, 2012, pp. 158-164. doi: 10.4236/ojg.2012.23016.

CHAPTER 6

Malanda, N. and Louzolo-Kimbembe, P. (2014) Survey on Storage of Water for Domestic Use in Underground Reinforced Concrete Tanks: The Case of Brazzaville (Republic of Congo). Open Journal of Civil Engineering, 4, 13-22. doi: 10.4236/ojce.2014.41002.

CHAPTER 7

P. Papadopoulos, H. Xenidis, P. Lazaridis, A. Diamantopoulos, P. Lambrou and Y. Arethas, "Achievements of Truss Models for Reinforced Concrete Structures," *Open Journal of Civil Engineering*, Vol. 2 No. 3, 2012, pp. 125-131. doi: 10.4236/ojce.2012.23018.

CHAPTER 8

Tosun, Y. (2014) Shale Stone and Fly Ash Landfill Use in Landslide Hazardous Area in Sirnak City with Foam Concrete. Geomaterials, 4, 141-150. doi: 10.4236/gm.2014.44014.

CHAPTER 9

M. Ghassemieh, M. Bahaari, S. Ghodratian and S. Nojoumi, "Improvement of Concrete Shear Wall Structures by Smart Materials," Open Journal of Civil Engineering, Vol. 2 No. 3, 2012, pp. 87-95. doi:10.4236/ojce.2012.23014.

CHAPTER 10

M. Ali, M. Alam, M. Haque and M. Alam, "Comparison of Design and Analysis of Concrete Gravity Dam," *Natural Resources*, Vol. 3 No. 1, 2012, pp. 18-28. doi: 10.4236/nr.2012.31004.

CHAPTER 11

Cheruiyot, J. , Abuodha, S. and Kabubo, C. (2014) Use of Stone Dust in the Design of High Performance Concrete.Open Journal of Civil Engineering, 4, 229-239. doi: 10.4236/ojce.2014.43020.

Index